T0231267

# LUBRICATION IN INLAND AND COASTAL WATER ACTIVITIES

# Lubrication in Inland and Coastal Water Activities

*"Growing acceptenace for biolubricants,
a benefit for the aquatic environment."*

*Edited by*

P. van Broekhuizen
*IVAM, Research & Consultancy on Sustainability, Dept. of Chemical Risks, Amsterdam, The Netherlands*

D. Theodori
*IVAM, Research & Consultancy on Sustainability, Dept. of Chemical Risks, Amsterdam, The Netherlands*

K. Le Blansch
*QA+, Questions Answers and More, The Hague, The Netherlands*

S. Ullmer
*University of Applied Science, Institute of Ship Operation,
Maritime Transport & Simulation, Hamburg, Germany*

A.A. BALKEMA PUBLISHERS   LISSE / ABINGDON / EXTON (PA) / TOKYO

The LLINCWA project was carried out with financial support from the European Commission, DG Enterprise, within the 5th Framework Programme in the Programme Innovation and SMEs, under contract number IPS-1999-0005.

Contributions from: Pieter van Broekhuizen, Demi Theodori, Hildo Krop, Cees van Oijen, Kees Le Blansch, Cees Ouwehand, Amaya Igarua, Philippe Lanore, Berard Lamy, Pascale de Caro, Christine Cecutti, Marie-Hélène Novak, Christian Busch, Uwe Schmidt, Holger Watter, Sylvia Ullmer

Copyright © 2003 IVAM UvA BV, Amsterdam, The Netherlands

*All rights reserved for LLINCWA partners. No part of this publication or the information contained herein may be reproduced, stored in a retrieval system, or transmitted in any form or by any means, electronic, mechanical, by photocopying, recording or otherwise, without written prior permission from the editors.*

*Although all care is taken to ensure the integrity and quality of this publication and the information herein, no responsibility is assumed by the publishers nor the author for any damage to property or persons as a result of operation or use of this publication and/or the information contained herein.*

Published by: A.A. Balkema, a member of Swets & Zeitlinger Publishers
www.balkema.nl and www.szp.swets.nl

Contact:
IVAM research and consultancy on sustainability
PO Box 18180, 1001 ZB Amsterdam, The Netherlands
tel: +31 20 5255080 – fax: +31 20 5255850
email: office@ivam.uva.nl – internet: www.ivam.uva.nl

ISBN 90 5809 612 2

Printed in the Netherlands

ISO Doc. Nr. 03-12-o

# Table of Contents

# Acknowledgment

Including the preparation of the project proposal and negotiations with the European Commission the LLINCWA project covered a period of nearly four years. A period in which a lot of political changes happened, merging of organisations took place, people found new jobs and last but not least the 6th Framework Programme was started.

LLINCWA was a successful project, exploring the lubricants world as a fertile soil to experiment with a technology transfer project trying to establish a sustainable substitution towards the use of biolubricants in inland and coastal water activities. LLINCWA was able to develop an enormous network of people and their organisations in one or the other way related to the use of lubricants.

Some of the organisations of the project partners turned out to be active merging companies during the LLINCWA-period, and in some of them many personal changes took place resulting in the fact that their successors had to take over their tasks, many times with a loss of build up expertise. These dynamic things seem to be normal in a modern society, but in a project like LLINCWA it puts an extra pressure on the responsible teams. Carrying out a large European project is not only a matter of carrying out the agreed tasks, but also a matter of forming a good, cooperative team. A team that is motivated to work together because they have been able to build a strong relationship. Therefore LLINCWA was never the reason for participants to change jobs. For most of them a promotion or a better fit to their personal needs could be realised with the new job. This foreword is the place to mention all the persons that participated in the LLINCWA team and to thank them for their contribution:

At the co-ordinator, the Chemiewinkel of the University of Amsterdam many personal changes took place followed by a merging towards an organisation under the Holding of University of Amsterdam: IVAM-Research and Consultancy on Sustainability, Department of Chemical Risks. Chronologically Maaike le Feber, Jaco Westra, Nicole de Boer, Cindy de Groot en Marianne Pieterse prematurely left the team. Especially the leaving of Jaco, as co-designer of the LLINCWA project meant a great loss. Two students Floris Hekster and Maria Frangou supported the team with specific projects.

Still going strong untill the end of LLINCWA are Pieter van Broekhuizen, Demi Theodori, Hildo Krop and Ckees van Oijen. For the financial administration of the project Annie Brouwer of IVAM and Jo Lansbergen of the Faculty of Science of the University of Amsterdam were the reliable beacons.

Non-changing and in full development during the LLINCWA activities was the group QA+ (Questions Answers and More) where Kees Le Blansch was the reliable partner.

And so was Cees Ouwehand from the waterboard Hoogheemraadschap van Rijnland. He only changed his working location but kept his position as permanent fighter for the biolubricants case, even in the UK trying to involve British Waterways in LLINCWA.

A special case is TotalFinaElf. Starting as Fina, having won an environmental price with biological screw axe lubrication products they were enthusiastic promoters to set up LLINCWA. A special word has to be given to Jacques Legrand who played an important role in the period before LLINCWA started. Without his support LLINCWA wouldn't have existed. His colleague Ives Mels who was thought to participate in LLINCWA already left after the first LLINCWA meeting, due to the merging of Total with Fina, to be succeeded by William Holtappels. Further merging with Elf forming TotalFinaElf resulted in a next, but this time sustainable change towards Philippe Lanore and Bernard Lamy.

Valonal and the University of Gembloux only faced personal changes in the very first beginning of LLINCWA. Ines Dufeÿ was succeeded by Marie-Hélène Novak. Nevertheless also Valonal faced the merging adventures and they formed the group ValBiom a group with a comparable mission as Valonal.

Our Spanish partner Tekniker had a more quiet approach towards the project. Amaya Igartua came and stayed and supplied the team with valuable tribological knowledge.

And so was the French team of the Institute Nationale Polytechnique de Toulouse where Pascale de Caro and Christine Cecutti were stable partners that supplied the team with important environmental knowledge.

Stable nearly until the LLINCWA end was also the German partner ISSUS, the University of Applied Sciences, Institute of Ship Operation, Maritime Transport and Simulation where Sylvia Ullmer and Holger Watter made an enormous effort to set LLINCWA on the German map as ship operation scientists, without an initial overdose on lubricants knowledge. Sylvia only left the very last month but not without finalising all her tasks.

Fuchs Lubritech that was stable as well in their LLINCWA participation with Christian Bush and Uwe Schmidt, with their backing vocals Jean-Luc Hurth and Rolf Luther. They have been a real support concerning the understanding of the technical aspects of biodegradable lubricants.

Not really in the LLINCWA team, but involved as subcontractor during the Sailing Campaign was Stichting Reinwater. Heinz Boskma as board facility manager and Kees van Drie as sailor on board of the Reinwater ship supported by different captains took care for a safe and successful trip on the Rhine towards Strasbourg and back.

Furthermore, on the national levels many persons were active and gave valuable support to the LLINCWA activities. In particular we have to mention members of the

National Advice Commissions (see chapter 6) and of course all the persons involved in the pilot projects (see chapter 7). We cannot mention their names without forgetting many others, so therefore we thank them all together.

And last, but not least, within the European Commission we like to thank Guido Haesen from DG Enterprise who stimulated our LLINCWA activities strongly and gave many valuable advises to even perform better as an Innovation project. He learned us how to do it!

The LLINCWA team will stop its activities, but efforts will be made to continue the activities to stimulate biolubricants within the $6^{th}$ or following framework programmes.


Pieter van Broekhuizen, on behalf of LLINCWA

Amsterdam, March 2003
Den Haag
Leiden
Paris
Gembloux
Eibar
Toulouse
Hamburg
Weilerbach

# Executive Summary

LLINCWA (acronym for the project title loss lubrication in inland and coastal water activities) was a technology transfer project under the EU 5th Framework programme Innovation and SME, that stimulated the use of biolubricants.

Three years of activities to stimulate the awareness on the existence of biolubricants, testing and demonstration of the performance of these products in applications on and around inland and coastal waters has made it indisputably clear that biolubricants are available for the majority of applications, that they have at least comparable and sometimes even a better technical performance than conventional lubricants. The largest drawback is their high purchase price, higher than conventional, mineral oil based lubricants.

In the lubricant market there is lack on an agreed definition for biolubricants. Efforts of LLINCWA to realise an agreement amongst all actors in the lubricants market on a classification system for biolubricants failed due to a conflict in established interests. As a compromise an operational set of criteria was accepted as minimum requirement for biolubricants, based on the end product. This minimum requirement is lower than existing ecolabels (White Swan, Blue Angel, which set their criteria for single components), but less strict standards like Vamil do comply.

---

**The LLINCWA minimum requirements**

1. Ultimate biodegradation within 28 days higher than 60% according to OECD 301 B,C,D,F or higher than 70% according to OECD A, E or a primary biodegradation within 28 days higher than 90% according to CEC-L-33-A-93

2. The acute aquatic toxicity (EC50/LC50) should be higher than 1 mg/l according to both OECD 201 and 202 or equivalent tests

3. No substances with R-phrases or combinations thereof in relation to sensitization, carcinogenity, mutagenicity and reproductive toxicity (R39, R40, R42, R43, R 45, R46, R48, R49, R60-R64)

4. No substances found on the EU list of priority substances in the field of water policy (COM(2000)47 final; Brussels 07.02.2000)

---

Lubricants based on natural oils (tri glycerides), synthetic esters and some poly glycols do fall within this definition.

It is concluded that there is strong need for the development of a harmonized definition for biolubricants. The development of a European ecolabel with strict criteria may play an important role in this respect.

The actual use of biolubricants in applications on and around fresh waters is extremely low. Market shares do not raise above 0,2 – 1% for bio grease and bio gear oils, somewhat higher for 2-stroke oils (2,8%) up till maximum 5% in some countries for bio hydraulic oils. The use in "aquatic applications" is not separately kept by lubricant suppliers, which makes it difficult to give a more exact estimation of the actual use.

Losses of lubricants due to inland and coastal water activities are significant. Of the totally 5 million tons of lubricants use on land and water almost 45 % is not recovered, meaning that they are somewhere lost in the environment. A comparable situation seems to take place in inland waters. Mineral oils do contribute to a significant amount to the total load of contaminants of the inland waters.

As a general statement one can say that the toxicity of (bio)lubricants is determined by the additives used, while the biodegradability is determined by the base fluids.

The main environmental problems with mineral oil used in lubricants are highlighted in its physical effect of staining essential organs of aquatic organisms and its low biodegradation (both aerobic and anaerobic). According to the EU criteria its acute aquatic toxicity is too low to classify many mineral oil distillates as hazardous for the (aquatic) environment. In contrast to mineral oil lubricants biolubricants, like vegetable oils, synthetic esters and some polyglycols show ready biodegradability under aerobic as well as anaerobic conditions. The toxicity of the oils and esters is low. Due to their ready biodegradability staining effects due to biolubricants are not likely to occur and long-term effects can be ruled out.

LLINCWA did set up many pilot projects, trials with lubricated equipment on and around inland waters in which conventional lubricants where substituted by biolubricants. They are described in more detail in chapter 7. Substitution took place in applications with loss lubrication and in system with a high risk for accidental loss in the environment (screw axe and rudder systems on inland ships, bridges, sluices, cranes, winches, gear boxes, etc). No structural objections were found that could limit the usability of biolubricants. For almost all applications biolubricants could be found and a proper substitution could be realised, which proofed biolubricants to be acceptable alternatives as a tool in preventing the pollution caused by conventional mineral oil based lubricants.

Nevertheless, despite all the positive signs of good technical performance, beneficial environmental behaviour, lower human toxicity and a relatively trouble-free substitution, the substitution towards biolubricants in practice finds many barriers in its way.

The high purchase price is identified as a main barrier that inhibits an overall successful substitution. An inventory of barriers shows a low market activity of suppliers to market specifically *bio*lubricants: low availability of specialised products, low recognisability, lack of supply in the right package size, lack of guarantee of OEMs for a safe use in their equipment. A continued emphasis to increase (but also keep) the awareness of the existence and environmental need to use biolubricants remains an essential need.

Voluntary measures alone, to stimulate the use of biolubricants are *not* thought to be enough to gain a significant market share, to reach a critical mass. Stimulating governmental measures are needed to overcome the high purchase price. Public procurement is thought to be an instrument that can further favour the use of biolubricants.

Finally it can be concluded that the chosen innovation approach for LLINCWA, possible within the *Innovation Programme*, with a strongly market oriented content focussing on the identification of drivers and barriers and trying to formulate integrated answers to all these obstacles, has resulted in a strong awareness about the existence and benefits of biolubricants in environmental *and* technical terms, and it was a stimulus for the industry to continue and even intensify its R&D and marketing activities to develop and sell biolubricants.

Now it's the national and European governments' turn as well to adopt the conclusions and transform them in national and European interventions, stimulating measures and legislation.

# PART 1

# GENERAL INTRODUCTION

# Chapter 1

# Introduction

## 1.1 INNOVATION PROGRAMME PHILOSOPHY

The leading concept in the EU Innovation programme is the conclusion that today the acceptance and integration of innovation matters is not only dependent on the new technical solutions they provide, but also on social, economical and environmental sustainability. Emphasis must be put on the non-technical aspects of the innovation process to realise a successful introduction of the innovative product. Therefore innovation projects make a combination of methodological *research elements* to produce new knowledge on economic, social and organisational aspects of innovation, with *demonstration elements* of technology transfer.

The research elements that explore the non-technical issues of the innovation process can be quite divers: novel approaches to training, management, co-ordination, communication, regulatory issues, exploitation and diffusion with the aim to obtain new knowledge, developing skills, products, processes or services, facilitating harmonisation between member states and exploring new channels for the diffusion of information. The demonstration elements are identification, adaptation, validation and documentation of technology transfer and as such include necessary implementation for the targeted user sector, with the aim at satisfying user needs.

The goal of the Innovation programme is to increase the public awareness on the existence of the innovative product, to make people acquainted and to realise its acceptance in the market.

## 1.2 LLINCWA

LLINCWA is the acronym for the Innovation project *Loss Lubricants in Inland and Coastal Water Activities.* It was carried out within the frame of the Innovation Programme of the EU 5th Framework Programme.

The general objectives of the LLINCWA project are: reduction of diffuse water pollution with lubricants and greases, increased use of non-toxic biodegradable lubricants, and protection of fresh water and the coastal zone. To achieve these goals important changes in the lubricant market have to be initiated and several substitution-oriented

objectives are formulated. Increase of the number of users to add to the critical mass needed for market acceptance of bio-lubricants, stimulation of self-organising and self-regulating processes on the lubricant market and increasing the transparency of the lubricant market and environmental methodologies.

The enormous scale in which lubricating agents are used both on land and in water motivated the development of the LLINCWA project. The high consumption volume contributes significantly to the diffuse contamination of both soil and surface water. That's why LLINCWA is focussed on prevention and diminishing pollution of the environmental compartment with the highest hazard potential, surface water. Although initially a restriction was made within the project to limit the activities to the use of *loss* lubricants, already at the start of the project in 1999 the scope was widened to include the hydraulic fluids and gear oils as well. Loss lubricants are formulated and used to be lost in the environment during use, subsequently being a direct environmental pollutant. Hydraulic fluids on the other hand are formulated for use in contained equipment, closed systems like pumps and hydraulic equipment. Nevertheless practice shows a significant environmental loss of these fluids due to leakages and breakage of tubes, making these fluids so called *lost* lubricants. Therefore LLINCWA deals with loss and lost lubricants.

LLINCWA activities took place in five countries, the Netherlands, Germany, Belgium, France and Spain, but during the project many contacts were made with other European countries.   Project partners were found among knowledge providers (universities and consultants), lubricants suppliers / manufacturers and users (water board), but an almost unlimited amount of contacts were settled in the lubrication world: scientists, manufacturers, suppliers, lubricants users, organisations of users, national governmental organisations, European organisations, original equipment manufacturers, agricultural (lobby) organisations, environmental organisations, trade unions, scientific and popular press etc.

To support the activities and to give LLINCWA a broad national scope all LLINCWA countries established an advice group, consisting of representatives of national and local governments (ministries of environment), lubricant supplier or their association, shipping organisations, technical organisations, and water quality management organisations. These advice groups took care for tuning strategic decisions made within the LLINCWA project into line with the whole working field.

Three main activities were carried out by the LLINCWA project:
- Research on technical and non-technical issues, to assure a reliable LLINCWA, to get a thorough understanding of the technical, the health and the environmental aspects concerning the aimed substitution of "traditional" lubricants with biolubricants.
- Setting up of pilot projects to study the problems of introducing biolubricants into equipment (the technical as well as non-technical problems of substitution) and to find out what barriers have to be taken in practice to get the biolubricants accepted by potential users.
- Dissemination of the gathered information to the market: to potential users and their organisations, suppliers and to relevant governmental authorities. Dissemination was carried out using newsletters, the production of informational material, the organisation of workshops and a conference, campaigns oriented at specific branch segments, discussions with branch organisations and organising pressure on governmental authorities to develop legislative initiatives to stimulate the use of biolubricants.

## 1.3   ACCOMPANYING MEASURES

A beneficial feature within the Innovation Programme was the existence of *accompanying measures (AM's)*. These were separately organised projects, financed by the same programme, and giving technical and management support to the specific innovation projects. Like the innovation projects the AM's were consortia of different organisations located in different countries with acronym names like: Pride, Ecoinnovation, CLIP, Lifestyle, Strategi'st, Showcase[1]. All were active in the Innovation Programme to provide specialised support and LLINCWA cooperated with most of them. Their support to the LLINCWA project is described in the different chapters, but mainly in chapter 7.

## 1.4   STRUCTURE OF THE REPORT

The different topics and results, including the conclusions of LLINCWA team concerning the potential for biolubricants to sweep a significant part of the lubricants market are described extensively in this report. The report is structured as follows.

The report contains five parts, *the first* being the general introduction into the innovation philosophy and LLINCWA's design.

*The second part, "The bio lubricant product",* gives a description of biolubricants, their technical and environmental properties. A description of the actual and potential market for biolubricants in the aquatic environment is given, attention is given to existing ecolables, indicating a challenge for complying biolubricants.

The third part, "Stimulating the market introduction of biolubricants", LLINCWA's approach to stimulate biolubricants. It gives an analysis of the market introduction seen from the perspective of the different user groups, the inland skippers, water management, recreational water users, the differences between the management and the technicians within a lubricant using company, the interests of suppliers and the role of governments in setting rules to take care for a clean environment.

*The fourth part, "LLINCWA's practical experiences",* forms the heart of the end report. It gives an extensive description of the pilot projects that were carried out during the three years of LLINCWA's activities, successes and failures are highlighted, technical details are given and an estimation is made of costs related to a substitution of mineral oil based non-biodegradable, toxic lubricants by biolubricants. Based on these experiences a manual is presented that may serve as a tool in substitution. This manual can be used by lubricant users to prevent failures and to take experienced barriers in future.

Furthermore LLINCWA's unique innovation approach is further elaborated, key points to explain successes and failures that were faced during the project are identified. These concern legitimising aspects, LLINCWA's cooperation with a heterogeneous market and their actors, the internal LLINCWA organisation and the way the project design helped to overcome scepticism. This part ends with the proceedings of the final LLINCWA conference.

---

[1] A description of the Accompanying Measures as well as of the Innovation Projects can be found on www.showcase.com

*Finally part five, " Future for biolubricants"* presents the final conclusions of LLINCWA. Based on the practical experiences and based on the market analysis an outline for the future for biolubricants is given. The need for further governmental initiatives, to guarantee a sustainable bio lubricant use in the aquatic environment, is emphasized.   Initiatives for further cooperation within the (bio)lubricant world are outlined.

# PART 2

# THE BIOLUBRICANT PRODUCT

# Chapter 2

# Description of biolubricants and their technical performance

## 2.1    INTRODUCTION

Most lubricating oils, greases and hydraulic fluids are currently obtained from distillation of crude petroleum. However due to growing environmental awareness and (the prospect of) stringent regulations regarding the use of petroleum products the manufacture and use of the so-called 'biolubricants', has began to gain importance.

There is as yet no universally accepted definition for biolubricants. Biolubricants for the purpose of this chapter will be those lubricants meeting the minimum LLINCWA criteria. These criteria involve a judgment about the biodegradability, the aquatic toxicity and the human toxicity of the end product as well as requirements regarding its technical performance.

This chapter describes the main characteristics and key issues related to the technical performance of biolubricants and the role of additives. Prior to the discussion concerning their technical performance, the LLINCWA definition of biolubricants is given.

## 2.2    WHAT IS A 'BIOLUBRICANT' ACCORDING TO LLINCWA

The term 'biolubricants' or 'biolubs' includes lubricating oils, lubricating greases, 2-stroke engine oils, hydraulic fluids and chain oils that are non-toxic to both humans and aquatic life and can biodegrade in relatively short time.

The requirements that LLINCWA employs in order to differentiate between acceptable and not acceptable products are derived from standards found in the European market.

The LLINCWA minimum requirements are as yet less rigorous than the criteria defined in the White Swan ecolabel (Nordic countries) or the Blue Angel ecolabel (Germany) and the Swedish Standards. They are most similar to the requirements of the VAMIL directive (The Netherlands). By setting the LLINCWA minimum criteria at the level of the VAMIL directive, LLINCWA has adopted a pragmatic approach that combines scientific analysis and the need to bring into line existing national policies on biolubricants.

The LLINCWA <u>minimum requirements</u> apply to the end product as follows:

| The LLINCWA minimum requirements |
|---|
| 1. Ultimate biodegradation within 28 days higher than 60% according to OECD 301 B,C,D,F or higher than 70% according to OECD A, E or a primary biodegradation within 28 days higher than 90% according to CEC-L-33-A-93 |
| 2. The acute aquatic toxicity (EC50/LC50) should be higher than 1 mg/l according to both OECD 201 and 202 or equivalent tests |
| 3. No substances with R-phrases or combinations thereof in relation to sensitization, carcinogenity, mutagenicity and reproductive toxicity (R39, R40, R42, R43, R 45, R46, R48, R49, R60-R64) |
| 4. No substances found on the EU list of priority substances in the field of water policy (COM(2000)47 final; Brussels 07.02.2000) |

Note that LLINCWA does not apply exclusion requirements with regard to the renewability of the product.

Note as well that these requirements are the minimum expected. In order to distinguish between biolubricants that demonstrate environmental performance over-and-above these minimum requirements LLINCWA has proposed a classification system. The LLINCWA classification system consists of scores I, II, III and IV for hydraulic fluids and scores I, II, III for gear oils, greases and lubricating oils corresponding to high, medium and low  - but still acceptable - environmental performance. The possible scores and their definitions are given in chapter 4 (Environmental and workers health aspects).

There are different types of base fluids in current commercial use that are marketed with the claim that they are biolubricants. However, only vegetable oils and a great number of synthetic esters easily meet the LLINCWA biodegradability criteria. This is also true for some low molecular weight polyglycols (PAGs). Analogous claims for polyalphaolefins (PAOs) are not supported by the data gathered within LLINCWA. The low viscosity types of PAOs of which it is claimed that they are ready biodegradable, are not generally used for the formulation of lubricants. PAOs are therefore not embraced by the LLINCWA definition of biolubricants and will subsequently not be dealt with in this chapter. Mineral oil base oil too does not satisfy the LLINCWA requirements.

Figure 2.1 shows the different types of base fluids frequently applied in biolubricants, which fall within the LLINCWA definition.

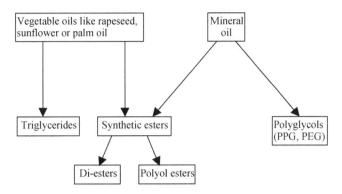

Fig. 2.1.   Sources of base fluids frequently encountered in biolubricants

The different features of biolubricants are described hereafter. For this purpose three main categories of biolubricants are distinguished related to the type of base fluid used to formulate the finished product; biolubricants based on vegetable oils, synthetic esters and polyglycols.

THE CLEAR ADVANTAGES

All biolubricants have a number of clear advantages over mineral oil based lubricants:
- high biodegradability
- low toxicity for humans and aquatic organisms
- good lubrication properties
- high viscosity index
- high flash point
- good adhesion to metal surfaces

## 2.3   BIOLUBRICANTS BASED ON VEGETABLE OILS

Vegetable oils consist of long chain saturated or unsaturated fatty acid triesters of glycerol. Types of vegetable oils utilized in the formulation of biolubricants include rapeseed, soybean, sunflower, palm, coconut, olive oil etc. Rapeseed oil is one of the most frequently used oils for biolubricants. The first rapeseed oil based fluids were commercially available already in 1985. A generic chemical structure is shown in Figure 2.2.

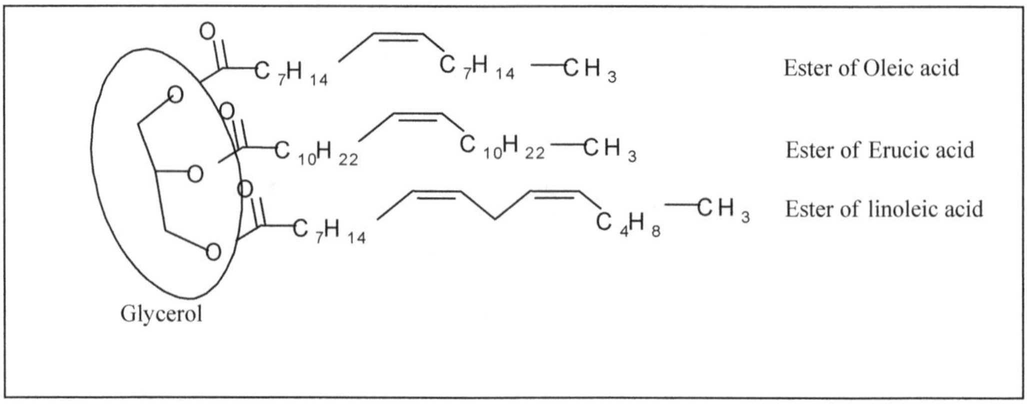

Fig. 2.2    Triglyceride of different unsaturated fatty acids and glycerol

Vegetable oils have some undeniable advantages. They have the environmental advantage of being non-toxic, ready biodegradable and renewable. In addition, they are lighter than water. Escaped vegetable oils can therefore be easily skimmed off the surface of water. Besides they have the economic advantage of being relatively inexpensive compared to other types of base fluids that can be used for the formulation of biolubricants.

*Performance advantages*

On the level of technical performance, vegetable oil based lubricants have the following advantages compared to mineral oils: good boundary lubricating properties, high viscosity index, high flash point, low volatility and better solvency for additives and products of oxidation. In addition, vegetable oils are fully mixable with mineral oils, and in most cases compatible with seal materials, varnish or paint although swelling of elastomers has also been reported.

Because of their higher polarity, vegetable oils have a higher affinity for metal surfaces than non-polar hydrocarbons, like mineral oils. Vegetable oil based lubricants provide therefore lower friction and less wear than mineral oils. The higher viscosity index allows the use of vegetable oils of lower viscosity classes than mineral oil based ones for a certain application. The utilization of low viscosity classes provides the additional advantage of easier heat transfer. These properties result in higher tool life.

*Performance concerns*

Performance drawbacks of lubricants based on vegetable oils include poor low-temperature fluidity, rapid oxidation especially at high temperatures and poor hydrolytic stability.

*Low-temperature fluidity*

Vegetable oils have high pour points, which limit their usable temperature range. Different triglycerids solidify and melt at different temperatures ranging from 74°C to -24 °C but most formulated products based on vegetable oils will solidify upon long-term exposure at circa -10°C. This temperature can decrease by means of additivation and also mixing with esters.

The fluidity of a material at low temperature is mainly depended on the efficiency of molecular packing, intermolecular interactions and molecular weight. In vegetable oils, the presence of double bonds has a positive effect on the low-temperature behavior of the oil. However, these same double bonds are responsible for the poor oxidative stability of the oil. The poor low temperature behavior of vegetable oil products can be partly improved by the use of additives (pour-point depressants).

*Oxidative stability*

The most serious disadvantage of vegetable oils when used in lubricants is their poor oxidative stability especially at elevated temperatures. Oxidation leads to polymerization, degradation of the lubricant and loss of functionality. When oil oxidizes it will produce acid and sludge. Sludge may settle in critical areas of the equipment and interfere with the lubrication and cooling functions of the fluid. The oxidized oil will also corrode the equipment.

The rate of oxidation depends on the degree of unsaturation, that is the number of double bonds, of the fatty acid chain. Bis-allylic hydrogens in methylene- interrupted polyunsaturated fatty acids present in linoleic and linolenic fatty acid are very susceptible to free radical attacks, peroxide formation and production of polar oxidation products. This oxidative susceptibility increases exponentially from an allylic methylene to a double allylic methylene. In general, the rate of oxidation of linoleic (18:2)[2] is ten times greater than that of oleic (18:1), while the oxidation rate of linolenic (18:3) is twice as vast as that of linoleic (18:2) fatty acid chain. The hydroperoxides thus formed undergo homolytic cleavage leading to polymerization. In addition, under thermal conditions the double bonds in polyunsaturated fatty acids isomerize to form conjugated fatty acids that polymerize much faster than their precursors.

Antioxidant additives improve oxidative stability of vegetable oils only to a limited extent and hydrogenation (and/or advanced plant breeding and genetic modification) is necessary to eliminate bis-allylic hydrogen in order to increase the oxidative stability of the oil. On the other hand, the elimination of bis-allylic hydrogen will lead to the detriment of the low temperature properties of the vegetable oil since its low temperature behaviour is exactly benefited by a higher degree of unsaturation. This demonstrates the conflict of attempting both good low temperature behavior and the best possible oxidative stability in a given vegetable oil based lubricant. Optimally, vegetable oils having high percentages of monounsaturated fatty acids offer the best compromise between high oxidative stability and good low temperature behavior. The oxidative stability of high oleic oils is three to six times greater than conventional vegetable oils. At present, high oleic acid sunflower oil (HOSO or HOAS) and / or low

---

[2] (carbon chain length : number of double bonds)

erucic acid rapeseed oil (LEAR) - both contain high percentages of oleic acid – are the most promising vegetable oils being developed for lubricants that can perform adequately in equipment performing in a temperature range between -15°C and 150°C. The maximum temperature depends on the distribution of triglicerides and on the additivation. Figure 2.3 shows the development of the low temperature behaviour and oxidative stability as a function of the degree of (un)saturation of different fatty acids.

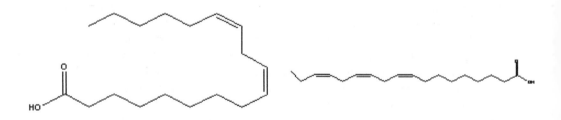

Stearic acid(18:0)
Iodine value: 0
Melting point (°C): + 74
Oxidation: low
Relative rate: 1

Oleic acid(18:1)
Iodine value: 86
Melting point (°C): +5
Oxidation: Moderate
Relative rate: 10

Linoleic acid (18:2)
Iodine value: 173
Melting point (°C): -11
Oxidation: high
Relative rate: 100

Linolenic (18:3)
Iodine value: 261
Melting point (°C): -24
Oxidation: very high
Relative rate: 200

Fig. 2.3 Low temperature behaviour and oxidative stability of fatty acids

*Hydrolytic stability*

As stated before, vegetable oils are esters of glycerol. Due to their ester structure, vegetable oils are prone to hydrolysis. However, the reaction between a pure ester and water is very slow. Vegetable oil based lubricants containing <1% water (w/w) can be stored at ambient temperature and undergo essentially no hydrolysis.

For hydrolysis to occur at a significant rate, some form of catalysis is required. Catalysts for ester hydrolysis are typically acids or bases. Additionally catalysis by metals and enzymes is also important. In practice only acid catalyzed hydrolysis is likely to occur. Acid-catalyzed hydrolysis is an autocatalytic reaction. Such reactions typically show a long induction time where little reactions occur; followed by onset once the level of catalyst (which is at the same time a reaction product) has increased to a sufficient level. The duration of the induction period depends mainly on the initial acid value of the product, and the concentration of any other species, which can react with the catalyst/product as it is produced (for example active metal surfaces). If the initial acid value is too high, then the induction period will be very short. The hydrolytic stability of vegetable oils can be therefore improved by good control of the initial acid value (mgr KOH/g) and only to a limited extend by the use of acid-catcher additives. In practice the hydrolysis does not need to be a problem if one cares for that the water content of the lubricant stays under acceptable limits. Finally, vegetable oil based lubricants are relatively more prone to hydrolysis than synthetic ester based lubricants due to steric-hindrance effects and the fact that vegetable oil based lubricants commercially available appear to have a rather elevated initial acid value.

Other areas of concern related to the use of vegetable oil based lubricants are the following:

- The results of the 4-ball tests show that the vegetable oils have good extreme pressure properties and anti-wear characteristics in relation to the mineral oil
- Vegetable based lubricants darken if exposed to light. Photosensitive lipids, fatty materials contained in oil, absorb UV light and change color. Although this may not change the lubricant's physical characteristics or performance, regular sampling and testing of the fluid may be necessary.
- Water and /or bacteria may enter the system and degrade the vegetable oil based lubricant. Given the right conditions, bacterial growth may cause color change, odor problems and loss of performance. In order to avoid this possible effect the water content should be lower than 1% (w/w).

## 2.4    BIOLUBRICANTS BASED ON SYNTHETIC ESTERS

There are two main types of synthetic ester base fluids in current commercial use: dibasic esters (diesters) and polyol esters.

Diesters[3] are derivatives of aliphatic diacids like adipic, azelaic, sebaic acid or other dicarboxylic acids esterified with monoalcohols. The most used raw materials are 2-ethylhexyl alcohol and azelaic acid, an acid derived by the ozonolytic cleavage of oleic acid.

Polyol esters are derivatives of neopentyl alcohols esterified with monoacids. This last group can be further divided into saturated esters and oleochemical esters.

Synthetic esters have some similarities with vegetable oils. They show good environmental performance in terms of toxicity and biodegradability. Other similarities involve good boundary lubricating properties, high viscosity index, high flash point, low volatility and good solvency for additives and products of oxidation. In addition, synthetic esters are fully mixable with mineral oils. While there are similarities between vegetable oil bsed lubricants and synthetic ester, there are also important differences. Differences with vegetable oils include higher purchase costs and incompatibility with some paints and seal materials. Additionally, synthetic esters score lower with regard to the renewability. The majority of synthetic esters utilized for the formulation of lubricants is only partly based on renewable materials while there are also synthetic esters which are entirely synthesized from mineral oil.

However, the most important differences between vegetable oil based lubricants and lubricants based on synthetic esters concern technical performance advantages of synthetic esters above vegetable oils: better low temperature fluidity, better oxidative and thermal stability and better hydrolytic stability. One the other hand, the cost of synthetic esters is much higher than those of mineral and vegetable oil.

*Low temperature fluidity*
Synthetic esters can be carefully designed in order to show good low temperature properties. This is achieved by the synthesis of molecules showing a high degree of asymmetry since symmetric molecules have the tendency not to crystallize, due to the absence of repeatable crystal lattice.

*Thermal and oxidative stability*
Thermal and oxidative decomposition are separated, though related, processes. In general, synthetic esters have good thermal stability compared to mineral oils. The main thermal decomposition pathway for synthetic esters is olefin elimination, which requires a hydrogen atom at the beta position of the alcohol group. Polyol esters, which have no beta hydrogen atoms, show consequently even better thermal stability than

---

[3]    Not to be confused with the French name "Diester" meaning the brand name of the biodiesel (rapeseed methyl ester) in France

diesters. The thermal decomposition for polyol esters is a free radical dissociation. Branching on the acid group close to the ester linkage in polyol esters causes steric crowding which increases the strain energy of the ester relative to the separated free radicals, so branched polyol esters have lower thermal stability than linear analogues.

Oxidative decomposition is a complex free-radical process, which occurs in the presence of air. The process is initiated by reaction of oxygen at the most reactive site of the ester. Where olefinic unsaturation is present (C=C bonds), there are the most reactive sites. For fully saturated molecules, reaction occurs at the C-H bonds. Tertiary C-H bonds are significantly more reactive than methylene or methyl groups. Branched molecules show therefore lower oxidative stability than linear molecules, except where the branching occurs at quaternary carbon atoms.

*Hydrolytic stability*
As stated before, synthetic ester based lubricants have better hydrolytic stability than vegetable oil based ones due to steric-hindrance effects.

Synthetic esters are designed for operating temperatures ranging from –25°C up to 200°C.

As has been stated already, the results of the 4-ball tests show that the synthetic ester oils have good extreme pressure properties and anti - wear characteristics in relation to the mineral oil

## 2.5 BIOLUBRICANTS BASED ON POLYGLYCOLS

Given the current industry understanding, lubricants based on polyglycols are considered as biolubricants. However, according to the LLINCWA criteria only the low molecular types appear to show acceptable biodegradability in order to serve as base oils for the formulation of biolubricants. Another environmental concern associated with polyglycols is their miscibility with water. Due to this water miscibility escaped polyglycols that end up in water cannot be skimmed off, a property that is conceived to be less favorable from environmental point of view.

Polyglycols are the first biodegradable oils on the market. The main technical advantage of these types of fluids is their good stability within a broad range of temperatures, from -45°C to 280°C. The main technical drawbacks of polyglycols are their incompatibility with mineral oils and paints, sealing and filter materials.

## 2.6 ADDITIVES

It is important to realize that finished products are typically blends of base fluids (that make up 70 to more than 90% of the product) with several additives. Additives are intended to improve the technical performance but have generally a negative effect on the environmental performance of the finished product since most additives currently used for the formulation of mineral oil based lubricants are toxic and non-biodegradable. The required quality and concentration of additives in lubricants depend upon the specific application and the nature of the base oil. As stated before, vegetable oils, synthetic esters and some polyglycols in general fulfill the requirements of biodegradability and toxicity posed by LLINCWA, whereas mineral oil does not. However, the additives necessary to achieve the required technical performance are often the limiting components.

The following classes of additives are frequently used:
- Antioxidants, to prevent oxidation of the base fluid and thickener. Oxidation or aging leads to discoloration, burnt odours, induction of corrosion and decrease of viscosity.
- Viscosity modifiers, to balance changes in viscosity of the base fluid and thickener owing to temperature changes.
- Pour point depressants, to prevent co-crystallization of paraffinic components in the base oil and the polymer chain.
- Detergents and dispersants, to keep oil-insoluble combustion products in suspension and to prevent agglomeration into solid particles.
- Antifoam Agents, to prevent the lubricant from foaming
- Demulsifiers and emulsifiers, to prevent the formation of water-in-oil emulsions. They are all surfactants.
- Antiwear and Extreme pressure. To prevent wear when the hydrodynamic lubrication has not yet build up or in the case of extreme stress and forces of the moving parts.
- Friction modifiers, to prevent stick-slip oscillations and noises by reducing frictional forces
- Corrosion inhibitors for ferrous metals, to protect the iron surface from the attack of oxygen, moisture, organic acids and other aggressive substances.
- Corrosion inhibitors non-ferrous metals, to protect the non-iron surface from the attack of oxygen, moisture, organic acids and other aggressive substances.

The next table shows an overview of the most important additives typically added to a mineral based oil lubricant.

| Additive | Generally used percentage in lubricants | Grease | Hydraulic fluid | Gear oil |
|---|---|---|---|---|
| Pour point depressant | 0.05 - 1% | - | 0.05 - 1% | 0.05 - 1% |
| Anti foaming | 0.1 - 0.5% | - | <0.1 | Max < 0.1% |
| Detergent/dispersant | 1-5% | - | | - |
| Anti-oxidants | 0.2-4% | 0.10-1.00% | 0.10-1.00% | 0.10-2.00 |
| EP and anti wear | <5% | 0.5 – 5% | 0.20-2.00% | <5% |
| Anti corrosion/rust | 0.01 - 2% | 0.1 – 5% | <1% | 1-4% |
| Biocides | 0.1% | - | 0.1% | - |
| | | | | |
| Total additives in weight % | | 2-5% | 1-3% | <8% |

Table 2.1

In the past, it was assumed that additives designed for mineral oil lubricants perform similarly when used with biolubricants. This led to the production of technically inferior biolubricants. Many of the industry's negative opinions of biolubricants stem from these early products. More recently, companies have started reformulating and/or designing additives specifically for biolubricants with successful results. Suitable additives for biolubricants should exhibit fast biodegradability and low toxicity.

Some companies use still today conventional additives derived from petrochemical sources for the formulation of biolubricants while others use additives specifically developed for biolubricants. There are at present additive companies that have developed purely plant-based additive packages. Rheine Chemie Rheinau GmbH, for example is a German additives manufacturer that has identified a pool of biodegradable additives; GEMTEK in the US produces a full line of total loss lubricants and hydraulic fluids using vegetable seed oil telemers, other seed esters, and rarified waxy substances as additives; TESSOL GmbH (Germany) also produces additives from plant sources.

Due to the fact that "bio base oils" show a better viscosity behavior than mineral oils, viscosity index improvers or friction modifiers are often not necessary when formulating biolubricants[4]. Similarly, the use of anticorrosion additives, like calcium sulfonates, succinic acid derivates or ashless sulfonates, is less necessary when formulating lubricants based on vegetable oils and synthetic esters because vegetable oils and synthetic esters have themselves good anti-wear characteristics. These good wear characteristics stem from their high degree of polarity that gives them the tendency to form physical bonds with metal surfaces.

*Thickening systems for biogreases*

When formulating biogreases conventional thickeners are currently used. These often include lithium stearate and calcium stearate and mixtures of lithium and calcium soaps. Both of these thickeners are considered to be good biodegradable. However, calcium soaps are more preferable when regarding health aspects.

## 2.7 CONCLUSIONS

With careful selection of materials, biolubricants can be formulated that optimally combine environmental and lubrication requirements. One way to produce those lubricants is to tailor made their molecular structure to get the expected properties. Another way is by additivation.

Not all biolubricants are the same. Within LLINCWA three principle classes of biolubricants have been distinguished based on the type of base fluid: vegetable oils, synthetic esters and polyglycols. There are also classes within classes and the characteristics and performances of a biolubricants can vary considerably from one chemical structure to another and from one product within one class to another.

---

[4] The modern agro-industry plays a crucial role in the improvement of vegetable oils for non food applications. The selection of varieties and developments in biotechnology (also without the use of GMO techniques), enable the development of new varieties of oilseed plants providing oils rich in erucic or oleic acids. In particular, high content of oleic acid improves the thermal stability and oxidative resistance of final products. Therefore, oleochemists have at their disposal a new generation of vegetable oils whose characteristics are particularly adapted to lubrication applications.

Selection of the correct base fluid for a particular application is important. Vegetable oils and (oleochemical) synthetic esters prove to have the best environmental performance. In general, vegetable oil based biolubricants perform well under moderate temperatures. Synthetic esters and polyglycols are designed to meet more severe operating conditions. The use of some synthetic esters and polyglycols may be restricted by reasons of material compatibility and costs.

Vegetable oils have excellent lubrication qualities and are nontoxic and biodegradable. They are made from renewable resources such as rapeseed, sunflower, and soybean, and are much less expensive than synthetic fluids. Their chemical structures are triglycerides in which a variety of saturated, monounsaturated or polyunsaturated fatty acids are esterified to a glycerol backbone. The physical properties of a vegetable oil depend on the nature of its fatty acid composition. Optimally, vegetable oils having high percentages of monounsaturated fatty acids offer the best compromise between high oxidative stability and good low temperature behavior. Vegetable oils tend to oxidize at temperature above 90 °C - although vegetable oils with high percentages of oleic acid do exist that resist oxidation up to 150°C. Also, vegetable oils have a limited low temperature capability (-15 °C). This significantly affects the outdoor applications where systems may sit for extended period at sub-zero temperatures. The low temperature properties can be improved by mixture with esters.

Synthetic esters, mainly based on dibasic esters (diesters) and polyol esters, are regarded as the best among the biodegradable base fluids. The biodegradability of these oils is comparable to vegetable oils and their lubrication properties are very similar to mineral oils. Besides, some of them are partly based on renewable resources. The advantages of these oils are excellent fluidity, and low temperature and aging stability. Because of these, they provide wide operational temperatures (-25 to 200 °C) and have long shelf and service lives. One the other hand, the cost of synthetic esters is much higher than those of mineral oil.

Polyglycols are preferred in very high and very low temperature conditions (-45 to 280°C). According to the LLINCWA criteria only the low molecular types appear to show acceptable biodegradability in order to serve as base oils for the formulation of biolubricants. Due to their water miscibility escaped polyglycols that end up in water cannot be skimmed off, a property that is conceived as less favorable from environmental point of view. The main technical drawbacks of polyglycols are their incompatibility with mineral oils and paints, sealing and filter materials.

Up until today, a limited number of additives prove to be suitable for use in biolubricants. More emphasis need to be placed on the development of ready biodegradable and non-toxic additives that are compatible with vegetable oils, synthetic esters and polyglycols. The development of environmentally compatible  additives is expected to open new opportunities for the use of biolubricants.

REFERENCES CHAPTER 2

Bartz, Wilfried J. (1998), *Lubricants and the Environment*. In: Tribology International, Vol. 31 (1998), nr. 1-3, p. 35-47

Boyde, S. (2000), *Hydrolic Stability of Synthetic Ester Lubricants*. In: J. of Synthetic Lubricants, Vol. 16 (2000), nr. 4, p. 297-312.

Department of the Army, US Army Corps of Engineers (1999*), Lubricants and Hydraulic Fluids: Engineering and Design.*

Dicken, T.W. (1994), *Biodegradable Greases*. In: Industrial Lubricants and Tribology, Vol. 54 (1994), nr. 3, p. 3-6

Kodali, Dharma R. (2002), *High Performance Ester Lubricants from Natural Oils*. In: Industrial Lubricants and Tribology, Vol. 54 (2002), nr. 4, p. 165-170

Magaroni, David (1999) *High Performance Biofluids: Natural or Synthetic?*. In: Industrial Lubricants and Tribology, Vol. 51 (1999), nr.4, p. 11

Shell. (s.a.), *Milieuvriendelijker smeermiddelen*.

Wilson, Bill (1998). *Lubricants and Functional Fluids from Renewable sources*. In: Industrial Lubricants and Tribology, Vol. 50 (1998), nr.1, p. 6-15.

# Chapter 3

# The inland and coastal waters market for (bio) lubricants

## 3.1    INTRODUCTION

This chapter deals with where, in what applications and to which extent biolubricants are used in inland and coastal waters activities.

The chapter is divided in six sections, as follows. After this introduction in section 3.1, in section 3.2 the activities are described that take place on and around inland and coastal waters and that require the use of lubricants. The following sections specifically focus on biolubricants. Subsequently it is discussed what is known about the use of biolubricants on and around the water (since often data are lacking) (section 3.3), and in what specific market segments (section 3.4) biolubricants are used more and less in inland and coastal water activities. Next, some relevant backgrounds are described which explain for some of the differences between the use of biolubricants in the different (particularly: LLINCWA-) countries and regions (section 2.5) The chapter is concluded in section 2.6 with a number of general observations following from the market overview that has been presented.

## 3.2    LUBRICANT USE ON AND AROUND THE WATER: WHERE AND WHAT?

### 3.2.1    The activities

As the project name already makes clear, LLINCWA is about los(s)(t) lubrication for *inland and coastal water activities*. Therefore, the project focuses on those activities involving the use of lubricants that are potentially harmful to inland and coastal *waters*. In concrete terms this means that LLINCWA addresses the use of lubricants in as diverse places as ships, harbours, wastewater treatment plants, locks, dams and bridges. On the other hand, the use of lubricants in factories, railways and motorways falls outside of LLINCWA's scope – with the exception of those activities in factories, rail- and motorways that pose a direct threat to the water quality, e.g. because of their natural location near harbours and ports.

With *inland water activities* reference is made to activities like sailing and goods transport over rivers and channels, water management (including bridges), hydroelectric power generation, and sometimes even embankment activities. Coastal water activities include activities by coast guards, port authorities and coastal fishers.

Consequently, offshore activities, ocean steaming and fishing in international waters are excluded.

### 3.2.2   The applications

The applications on and around the water that require the use of lubricants are manifold.

- They concern for instance the propelling system of ships (motors, screw axis), and the machines and apparatuses on ships required to exert mechanical force (winches, cranes, et cetera). Also in steering and transmission mechanisms lubricants are used (e.g. as hydraulic fluids).
- In water management, gearboxes, hydraulic systems and underwater bearings for bridges, locks and sluices, as well as pumps and generators require lubrication.
- In hydroelectric power stations, motor oil for generators as well as hydraulic fluids are used.

Next to these generic applications, many highly specific applications exist on and around the water that require lubricants, like:

- Switch plates for crane rails in harbours
- Ship elevation systems near dams and in harbours
- Groundwork machines for river dams and watersides.

One way to classify all possible applications on and around the water that is used in the LLINCWA project, is to distinguish between:
- Floating equipment: ships (leisure, trade, maintenance, police)
- On-board equipment: e.g. cranes, winches
- Fixed equipment: locks, dams, power stations (hydroelectric), wastewater treatment plants (WWTP)
- Mobile equipment: e.g. cranes, harbour activities

### 3.2.3  The types of lubricants

Finally, the activities that LLINCWA focuses on can also be distinguished with respect to the types of lubricants that are used:

- *Greases*, used in typical loss-lubrication applications and also in closed systems
- *2-Stroke oils*, which are added to the fuel of 2-stroke engines and are burnt together with the fuel. The remains are emitted as part of the exhaust fumes into the air and are partly deposited into the water.
- *Gear oils and Hydraulic oils*, almost always used in closed application. Nevertheless serious percentages of these oils end up in the environment due to sweating, leaks and accidents. As was explained in chapter 2 of this report, this accounts for the inclusion of gear oils as well as hydraulic oils in LLINCWA.

All in all, a matrix can be presented for the different activities and applications and the types of lubricants they require.

Table 3.1:  Examples of different types of aplications and types of lubrication

|  | *2-Stroke oils* | *Greases* | *Gear oils* | *Hydraulic oils* |
|---|---|---|---|---|
| Floating equipment | Motorboats with 2-stroke engines | Screw axis | Gear boxes | Ship transmission systems, dredgers |
| On-board equipment | n.a. | Cranes, winches | Cranes, winches, gear boxes |  |
| Fixed equipment | n.a. | Locks, dams | Generators in hydro-electric power plants, gear boxes | Lock doors |
| Mobile equipment | Mobile equipment with 2-stroke engines | Cranes | Gear boxes | Mobile equipment transmission systems, excavators |

### 3.3  BIOLUBRICANTRICANTS USE ON AND AROUND THE WATER – THE AVAILABILITY OF DATA

In order to obtain insight in where, in what applications and to which extent biolubricants (further: biolubricants) are used in inland and coastal waters activities, the LLINCWA team has been keeping record of all available and accessible relevant information. In a structured attempt to obtain data the different national LLINCWA teams have been asking questions to suppliers, have studied various different types of market studies, reanalysed available statistics and sometimes carried out own research. Also, interviews were held with a number of so-called Original Equipment Manufacturers (OEM's). LLINCWA partner company Fuchs has shared its insights into the German biolubricants market. TotalFinaElf has done the same for the Dutch market, and has additionally provided us with information derived from their access to insights from a market study carried out by Frost and Sullivan (1999). Further contacts and exchange of information have been established with the European organisations Concawe (oil industry) and Fediol (vegetable oil producers).

The first conclusion that follows from our exploration (preceding the conclusions presented in the following sections) is that hardly any data are available or accessible to our teams that provide detailed and reliable insight into the use of bio lubricants for inland and coastal water activities. Apparently, LLINCWA deals with a difficult combination of categories (bio-lubrication, inland and coastal water activities), on which there are hardly serious independent market statistics. Suppliers indicate that they have limited insight into the precise application for which lubricants and biolubricants are purchased, particularly in cases in which products are sold through resellers.

It should therefore be understood that the data presented below are mostly based on estimates of market players and on other sources that are often – on their own – of questionable robustness. However, in combination the figures and data all tend to corroborate a number of general conclusions. These are presented in the following sections.

## 3.4    BIOLUBRICANT USE ON AND AROUND THE WATER – WHICH MARKET SHARES?

### 3.4.1   Introduction – the market of biolubricants in general

The overall conclusion following from the available data is that the use of biolubricants for inland and coastal water activities is easily overestimated. Substantial market shares of biolubricants for use in chainsaw oils and as hydraulic oils in forestry equipment tend to paint a modestly positive general picture that does not necessarily hold true for biolubricants in and around the water.

In 1999 the total European market for biolubricants was approx. 1,9% of the total lubricant market (0,1 of 5.4 million tons). In France this reportedly amounts to somewhere between 1,5% and 3%; in Germany this is between 4% and 5%; in Spain an estimate of 1% is mentioned. In Belgium calculations lead to the conclusion[5] that biolubricants have 0,05% of the market of lubricants, and 0,3% of the market for loss lubricants.

Over 60% of the biolubricants market concerns hydraulic oils (in France 3% of the total hydraulic fluid market; in Germany 11%; in the Netherlands this figure rose between 1995 and 1999 from 0,6% to 3,6%).

Next, 15% of the biolubricants market concerns chainsaw oils (in Germany 34% of chain saw oils are biolubricants), 8% concrete release agents (in Germany 87% of these agents are so-called 'VERA's') and the rest (amounting to over 10%) concerns other products (gear oils, 2-stroke engine oils (in France about 5% of the total market)), cutting oils and other).

---

[5] See: http://www.ienica.net/reports/belgium.pdf

### 3.4.2   Three main fields

Considering the different categories outlined in section 3.2, we will distinguish here between three specific types of biolubricants for three specific types of uses on which statements may be made in more general terms:
- Greases and gear oils for floating and on-board equipment
- 2-Stroke oils for outboard engines
- Hydraulic oils for on-board and fixed equipment (locks, sluices, wastewater treatment plants).

We will deal with these in the following three paragraphs.

### 3.4.3   Greases and gear oils

At this moment there are strong indications that biodegradable non-toxic greases and gear oils are hardly used for floating and on-board equipment. Market share estimates for the Belgian market amount to 0,0 percent, for the Dutch market between 0,2 and 1 percent,[6] for Germany 0,18 percent.

### 3.4.4   2-Stroke oils

As to the use of biodegradable non-toxic 2-stroke oils for outboard boat engines no specific data are available. We do have some data on the market share of these (biolubricant) oils in general (2,8%). Next to that, we know of legal regulations in several German *Länder* (as well as in non-LLINCWA countries Austria and Switzerland) which forbid the use of all toxic non-biodegradable products on and around lakes (particularly Lake Konstanz (since 1982) and the Bodensee, primarily in order to protect drinking water quality), which serves as more or less circumstantial evidence for the use of these types of 2-stroke oils for inland water activities.[7]

[6]   The first of these estimates is based on volumes, the second one is based on the number of Dutch ships using bio-products (in 2002 48 out of 10.000 vessels in the Netherlands (half of which are inland freight ships) use biolubricants).

[7]   The way it seems, however, the market shares that are reported concern biolubricants excluding synthetic products. It is claimed that meanwhile (*and mainly for technical reasons)* most 2-stroke oils are synthetic (as a consequence of which in Germany 90% of 2-stroke engine oils are said to be biodegradable).

### 3.4.5   Hydraulic fluids

Biolubricants hold the relatively largest market share in applications for hydraulic fluids (they generate up to 70% of total revenues on biolubricants in Europe). This appears to be partly related to the previously mentioned regulations, as well as to similar regulations for ecologically sensitive forest areas (and to relatively intensive ecological management initiatives in forest management in the Scandinavian countries). Moreover, technical and cost advantages of these bio hydraulic fluids appear to have been important driving forces behind their market shares.

As to the use of the fluids in inland and coastal water activities, precise data are lacking. The fact that these products are used for a widespread range of applications (e.g. certain types of transport vehicles, Schotteltype screws, mobile lifting equipment, hydroelectric power stations, lock systems) is an extra complication in attempts to obtain an overview.

### 3.4.6   Niches

Next, some specific niches can be mentioned of which we know that the use of biolubricants for inland and coastal water activities is relatively high:

- *Public procurement in Germany*
  Biolubricants are used more often on publicly owned vessels and installations in several German *Länder*, as a consequence of a call from the Federal Parliament to all public bodies to switch to biolubricants. Also, a call was made to report regularly on the progress made in substituting towards biolubricants. In this respect it was reported that public users applied 4% biolubricants for loss lubricants and greases, and 31% for hydraulic fluids. During the course of the LLINCWA project these market shares have reportedly increased further, which can be interpreted as a sign that critical mass has been reached. Some statements are made that these levels of biolubricant-use are already approaching saturation. Simultaneously, however, it appears that German municipalities only buy 1% biodegradable lubricants for their police and pawning boats.

- *Regulation on German Lakes*
  As mentioned before, on and around certain lakes and ecologically sensitive areas in several German *Länder* biolubricants are used more frequently, as a consequence of regulations that are installed to protect drinking water quality.

- *Dutch environmental policy*
  In the Netherlands there are indications that greases, gear oils and hydraulic fluids in publicly owned fixed equipment for sluices, locks and water treatment plants are relatively often of biolubricant quality. The extra attention from Dutch environmental authorities appears to be an important driving factor behind this.

## 3.5    BIOLUBRICANT USE ON AND AROUND THE WATER –
THE CIRCUMSTANCES IN DIFFERENT LLINCWA COUNTRIES

### 3.5.1    Introduction

Certain national characteristics or circumstances can be mentioned, which are of specific influence to the introduction of biolubricants in the different LLINCWA countries. Below they are briefly indicated for each country (in alphabetical order).

### 3.5.2    Belgium

Belgium presents a good example of both the trans-national and the country-specific nature of the LLINCWA issue. The trans-national aspect is a consequence of the close connection of the Belgian to both Dutch, French and German waterways, and is for instance highlighted by the fact that approximately 60% of the boats on Belgian waters are of Dutch origin.[8] Dutch regulation and quality standards are of particular influence to what happens on the Belgian inland waters, although there are also indications that Dutch boats alter their behaviour when sailing the Belgian waters.

A typically Belgian complication concerns the administrative division in Belgium, as a consequence of which three regions are competent for inland waterways.

An important factor in Belgium hindering the substitution towards biolubricants is the absence of a waste collection system. In the eyes of many parties concerned (amongst whom the skippers themselves), this problem outweighs and overshadows all other environmental issues connected to lubricant use for inland water activities.

It is mentioned that in Belgium a law is adopted that requires biolubricants to be used in all operations taking place near non-navigable waters. It is fully unclear, and doubtful, whether there is any effect to be perceived from such a law in Belgium on the use of lubricants for inland and coastal water activities.

### 3.5.3    France

In France both regulatory and market actors appear to be less active than those in other countries where the promotion of the use of biolubricants is concerned. Also, users appear to be less environmentally aware than in some other countries. In recent years slightly positive trends can be perceived in the interest for biolubricants of French public users.

---

[8] NL ships run 30% of kilometres counted in Belgian waterways each year. NL ships provide more than half of the transport capacity.

### 3.5.4   Germany

Germany is the largest lubricant consuming nation in Europe. In Germany both environmental and (lubricant) quality awareness appears to be highest of all LLINCWA countries. Also, the largest proportion of legal regulations and other initiatives (like the public procurement initiative and the eco-label *Blaue Engel* that is also applicable to biolubricants) is in place. Agricultural policy in combination with the stimulation of renewable raw materials has led to the FNR initiative for financial support of the use of bio-products. This also shows and explains Germany's leading position within the EU where the use of biolubricants is concerned.

As in Belgium, also in Germany it is reported that not just German boats sail the German waterways. It is mentioned that inland shipping in Germany is carried out by German ships (40%) and ships from other West-European countries (60%).

### 3.5.5   The Netherlands

In the Netherlands the active stance of the Environmental Ministry is a remarkable feature, resulting a/o. in the inclusion of biolubricants in equipment criteria for the VAMIL regulation (a fiscal measure allowing for quicker depreciation of environmentally sound investments) and in the 'BOMS'[9] consultation structure. Environmental awareness in the Netherlands is relatively high. There was also a Dutch ecolabel *('Milieukeur')* for lubricants, but since this was highly unsuccessful (no single company applied) this ecolabel was withdrawn.

---

[9]  BOMS = Beleidsoverleg Milieuvriendelijke Smeermiddelen, Policy Board on Environmental Friendly Lubricants (Ministry of Environment + Lubricants supplier)

### 3.5.6  Spain

In Spain the situation appears to be the same as in France. The fact that drinking water is generally scarcer in Spain than in other EU countries does not seem to have any impact on the use of lubricants (probably because many water activities in Spain happen in coastal waters).

## 3.6    CONCLUSIONS

In the previous sections available data have been presented concerning the use of biolubricants on and around the water. They are summarised in Table 3.2 below.

Table 3.2: Overview of market data

| | Biolub market share INCWA | Niches | Biolub market share Overall |
|---|---|---|---|
| Greases / oils | BE: 0,0%; DE: 0,18%; NL 0,2 - 1% | DE: • publicly owned vessels and installations:   greases/oils 4%   hydraulics: 31% • activities on and around lakes | |
| 2-stroke | ? (synthetics DE: 90%) | | FR: 5% |
| Hydraulic fluids | ? | NL: • publicly owned installations | DE: 11%; FR: 3%; NL: 3,6% |

The obvious conclusion is that the introduction of biolubricants on and around the water is still in a very early stage. There are signals and impressions that market shares have increased gradually and slowly, partly also due to LLINCWA's efforts. Among suppliers and users interest in and awareness of both the environmental relevance of lubrication and the availability of alternatives to mineral oil based lubricants have clearly risen. However the major part of the market still remains to be conquered.

On the one hand this underlines the importance of the LLINCWA project, as it is clear that a lot needs to be done in order to further the use of biolubricants in inland and coastal water activities. On the other hand this situation needs to be explained, in order to be able to develop effective biolubricant marketing strategies for LLINCWA and thereafter. Some explanations have already been hinted at in the description of some relevant national circumstances. We will turn back to this in chapter 5 of this report, in order to present an explanation for the low level of biolubricant use on and around the water in the LLINCWA countries.

# Chapter 4

# Environmental and health aspects

Traditional mineral oil based lubricants have significant drawbacks with respect to the environment and health hazards for professional users and consumers. Obviously biolubricants are not entirely harmless either. However there is a clear difference in the magnitude of the health and specifically the environmental hazards connected to mineral oil based products versus biolubricants. In this chapter the environmental issues will be dealt with first followed by a short description of the health hazards.

## 4.1    INTRODUCTION

Lubricants are essential in many applications and are consumed in high volumes. In Europe in 1999 the total lubricant consumption (on land and water) was almost 5 million tonnes. Owing to the presence of (waste) collection systems around 32% of the total amount could be recollected. An estimated 45% was, however, lost during use and 23% was unaccounted for. Large amounts are lost in to the environment, mainly into soil and water systems. Some part is lost deliberately for example, in order to prevent water from entering into the application. Examples are stern tubes, water locks, bridge systems etc. An estimated 300 tonnes of grease used in stern tubes is lost annually in the Netherlands and another 100 tonnes in Belgium only. Another part of lubricants is lost due to breaks, circuit leaks and "sweating" of closed (hydraulic) systems. Owing to the continuous loss from many different applications near or in water systems, lubricants are a source of (diffuse) water pollution.

Nearly all lubricants are based on mineral oil. It is a complex mixture of aliphatic, cyclo aliphatic and aromatic hydrocarbons. Owing to its physical, chemical and (eco) toxicological properties and the high volumes lost into environmental compartments, like water, mineral oil makes up an important chemical pollutant in water. In fact in the Netherlands (1) mineral oil is the second chemical pollutant encountered in the River Rhine after adsorbed organic halogens (AOX) (see figure 4.1). It is expected that this figure will not differ very much in other European waters.

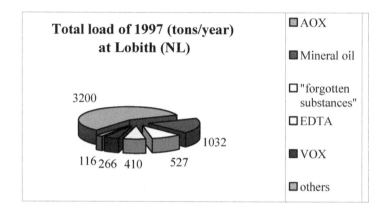

**Total load of 1997 (tons/year) at Lobith (NL)**

☐ AOX

■ Mineral oil

☐ "forgotten substances"

☐ EDTA

■ VOX

☐ others

3200

1032

116 266  410    527

Fig. 4.1. Relative proportions of total load of 1997 at Lobith, NL

Abbreviations: AOX -Adsorbed organic halogen, EDTA- ethylene diamine tetra acetic acid,
VOX-Volatile organohalogen compounds, Others PAKs, PCB's, perticides
Source: "Forgotten substances in Dutch surface waters", RIZA report 2001.020.

Loss lubrication represents more than 10% of total lubricant consumption. In addition between 20% and 30% of hydraulic fluids, 2-stroke engine oils, metal working fluids are also lost into the environment. In fact, between 20% (in EC countries) and 32% (in USA) of used lubricants are lost in the environment, including 15% in water.

Fresh water reserves on earth are limited and are essential to life not only owing to their use as drinking water for men and animal, but also to, e.g., the recreational or the fishing sector. It is known that small amounts of mineral oils stain both fish and drinking water; each litre of waste oil released into water can contaminate 1 million litres of drinking water. Mineral oil is thus one of the most important chemical pollutants in the European waters. Lubricants contribute substantially to this pollution, although it is difficult to estimate its contribution in this respect.

All lubricants contain, apart from their base fluids, a variety of other substances, like thickeners in greases and additives, small amounts of substances added to the lubricant in order to improve the lubricating properties. While nearly all attention is given to the base fluid (generally mineral oil), the environmental (and health) effects of these other added substances is much more difficult to assess. A recent assessment of the Dutch institute for Inland Water Management and Waste Water Treatment (RIZA) (1) on water pollution by chemicals identified around 150 « forgotten » substances in the Dutch fresh water systems. Normally these forgotten substances are not routinely analysed. Some of them could be linked to additives used in hydraulic fluids and were ranked high by using the same method the EU used to set up their priority substance list for the aquatic policy.

We may conclude that present day lubricants, based on mineral oil, are quite an important source of water pollution. To diminish its environmental impact, research into new types of lubricants that technically functions well but has much less impact on the environment has increased tremendously. It is, however, necessary to define what aspects specifically need to be improved in this respect.

While most attention to mineral oil based lubricants is given to the environmental aspects, its health effects cannot be ignored either. Therefore in this chapter the first section is devoted to environmental aspects, while in the second part its health aspects are considered.

## 4.2    ENVIRONMENTAL FATE

The environmental fate of mineral oil is determined by a sequence of abiotic and biotic processes, by the physical and chemical characteristics of the spilled oil and by environmental conditions, such as temperature, hydrodynamics and wind speed. The first process immediately following to a spill event is the rapid spreading of oil and drift, influenced by prevailing wind and current conditions. Both thickness and spreading-area of the spill are determined by factors such as viscosity, surface tension of the mineral oil and initial volume. Due to evaporation (especially of low boiling aliphatic and aromatic hydrocarbons, generally <u>not</u> abundantly used in lubricants) and to a lesser extent to solubilisation and formation of oil-in-water emulsions, the chemical composition and physical characteristics such as viscosity and surface tension of the oil undergoes rapid changes. In the next stage advection of oil films occurs under the influence of wind and currents. Drift of oil films usually proceed at 3-5% of the wind speed. Wind action may result in spatial inhomogeniety of spilled oil films, with a patchy-like distribution. (23)

Sorption to particulate matter or dissolved organic matter usually results in a limited bio-availability of hydrocarbons in the water column. Sedimentation of particulate-bound hydrocarbons results in accumulation of oil in sediment layers. Ingestion of particulate matter and dispersed oil droplets by zooplankton may result in aggregation into faecal pellets and enhance sedimentation of adsorbed hydrocarbons. Flocculation of oil droplets and coagulation with suspended matter may enhance transport of hydrocarbons to the sediment. (23)

In the assessment of the fate of biolubricants in the aquatic environment the behaviour of the lubricants (base fluids) under aerobic as well as anaerobic conditions is therefore inevitable.

## 4.3    WATER POLLUTION

Modern ecological assessment methodologies do use chemical as well biological analyses to assess the amount and extend of water pollution. In biological assessments one measures adverse effects on organisms like fish and daphnia's. Nowadays biological systems are used in many waters as warning systems or are assessed in so-called bioassays.

With chemical analysis one does assess the pollution of water by chemicals that may cause adverse effects on biological organisms. Nevertheless, the analysed chemical pollutants can explain only 20% of the assessed toxicity. These analyses are even complicated because the analysed waters may contain dissolved organic carbon (DOC), suspended particles and sediment remnants that all play a specific role in the ecotoxicological effects. These undefined particles can for example absorb part of the substances, which make these substances most likely not available for aquatic organisms. On the other hand chemicals can also enter the food webs and as a result accumulate in higher organisms.

Around 30 years ago some large negative effects of chemical pollution on water organisms were identified. One effect was the considerable decrease in size of colonies of specific aquatic organisms and birds depending on fish consumption. As a technical solution wastewater treatment plants were constructed, purifying the polluted waters from households and industries. As a result these acute lethal effects diminished considerably in Western countries. Also risk assessments for new substances entering the market were introduced to assess in advance possible negative (environmental) effects. These risk assessments are also used to calculate risks of substances still present in the environment.

Risk assessment of any chemical compound is based on two aspects:
I.   The amount or rather concentration within any organism that may cause a specific toxicological effect or in other words the maximum allowable concentration within an organism without any toxicological effect.
II.  The environmental activity (concentration) of this chemical compound (within the vicinity of the organism) low enough to prevent the toxicological effect mentioned in the first point.

To answer the first point standard (eco)toxicological tests have been developed measuring toxicological effects in specific organisms, organs or cells. Nowadays the results of these tests are translated into predicted no-effect concentrations in water (PNECs) or no-observed adverse effect levels (NOAELs) in terrestic organisms or humans.

In general very high precision cannot be achieved in ecotoxicological tests due to the naturally existing variation within populations. Standardisation can diminish this inaccuracy, but the result remains still arbitrarily since many assumptions are involved.

For the aquatic system standards are being used to regulate maximum environmental concentrations, Maximum Tolerable Concentration (MTC) or Maximum Tolerable Risk-level (MTR). The situation is often complicated since different methods are used to calculate these tolerable concentrations.

Answering the second point it is necessary to know emission locations and quantities, as well as the physical and chemical behaviour of the chemical in the environment on its way to the targeted organisms. Such behaviour for a chemical in water depends for example on volatization and the ad/desorption behaviour of the chemical to sediment or suspended particles and its transport speed that are all physical processes. The targeted substance may also be subjected to chemical changes like biodegradation or by sunlight.

The full behaviour is generally modelled in so-called fate models. These models and environmentally measured concentrations establish the predicted environmental concentration (PEC) in the vicinity of the targeted organisms or intake values by higher organisms.

The combination of the mentioned points I and II lead in water to a PEC/PNEC ratio for the specific chemical and organism. If the PEC/PNEC-ratio is higher than one, the substance is thought to cause adverse effects in the environmental compartment like water or in humans. A model that estimates both the fate (II) and the toxicological risk (I) is called a risk assessment program or model.

The EU is using the model EUSES to estimate these ratios and the underlying methods are described in the Technical Guidance Document 1488/94 and its updates. However, most of the used models are only applicable to non-charged organic substances. In this risk assessment the predicted no-effect concentration is obtained from ecotoxicological properties of the substance on organisms, while the (predicted) environmental concentration is calculated from exposure scenarios, loss and uptake due to physical or chemical processes like biodegradation, hydrolysis, photolysis and bioaccumulation.

Quite a number of these properties are specific for the chemical itself and are used to establish a hazard classification. Well known is the EU classification on substances and preparations as laid down in Directive 67/548/EEC and 1999/45/EU (laws, regulations and administrative provisions relating to the classification, packaging and labelling of dangerous preparations) and updates. In addition national initiatives in a number of EU countries are also developed like in Germany the Water Hazardous Classification system (WGK); one of the most well known systems for the water compartment. Any classification system is based on the result of number of standard laboratory tests. The EU has adopted a number of standard tests on health and environmental aspects. At this moment environmental aspects for products like lubricants cover only the water and ozone compartment in the EU. For water they comprise tests on biodegradation, bioaccumulation and ecotoxicity. Improvement on these environmental aspects is paramount in the development of a new class of lubricants, the so-called biolubricants. However, especially biodegradation and ecotoxicity are by no means uniquely defined and are therefore highlighted in the following sections.

## 4.4    ENVIRONMENTAL LOAD FROM LUBRICANT USE DUE TO INLAND AND RECREATIONAL SHIPPING ACTIVITIES IN THE NETHERLANDS

The three basic sources of mineral oil pollution due to lubricants use in the Dutch waterways are water management, inland shipping and recreational boating activities.

Within LLINCWA a study has been conducted in order to determine the volume and environmental significance of the emissions of mineral oil to inland and costal waters arising from lubricating activities of inland shipping and recreational boating in the Netherlands. Additionally the relevance of these emissions in the context of the total amount of mineral oil measured in the Dutch waterways has been estimated. Information concerning the lubricant releases has been generated through estimation methods and literature study (26,27,28,29). Analogous efforts aimed at estimating the amount oil releases from the Dutch Water Boards proved unsuccessful due to the lack of data.

### 4.4.1   Sources of mineral oil pollution for inland shipping

The findings with respect to inland shipping are mainly based on a study conducted by the Dutch Institute for Inland Water Management and Waste Water Treatment (RIZA) (27,28,29)[10].

Inland shipping activity may contribute to water pollution through the emission of lubricants that arise from the normal, operational running of inland shipping vessels, which are classified as 'operational discharges'. These include wire rope lubricants and other auxiliary machinery operating on the principle of loss lubrication as well as lubricating greases used in (semi) open stern tube and rudder trunk systems. The contribution of wire rope lubricants and auxiliary machinery lubrication is considered to be of minor importance when compared to stern tube and rudder trunk lubrication. Furthermore, possible discharges from closed stern tubes and rudder trunks are not expected to exhibit significant lubricant discharges under normal operating conditions, other than when an accident occurs, for which reliable data are hard to be obtained. This holds equally true for all kind of 'accidental discharges' due to failed hydraulic equipment or operational errors.

Additionally, lubricants may end up in the surface water through illegal discharges of bilge water. Bilge water is the wastewater generated in the bilge of the ship's machinery spaces. Not only leaching lubricants but also fuel contributes to the oily contamination of bilge water. The last years the pumping of bilge water into the waterways is prohibited for professional inland shipping vessels and skippers have to deliver bilge water to dedicated collectors. However, it is believed that still today a considerable amount of bilge water is illegally discharged into the waterways. In the year 2000 more than 300 bilge water spills with differing volumes (50 - >1.000 liters) were registered in the Dutch inland waters (25).

*Estimation of mineral oil discharges*

Consequently, for the estimation of lubricant emissions arising from inland shipping only the (semi) open grease lubricated stern tube and rudder trunk systems were taken into account. Vessels with such a (semi) open system make up the largest part (ca. 80%) of the Dutch professional inland shipping fleet, which in 2000 comprised ca. 4.571 active vessels.

A fraction (up to 40%) of the grease used for the lubrication of the stern tube and rudder trunk abandons the (semi) open system on the engine room site where it can be

---

[10]   RIZA is the research and advisory body for the Directorate-General for Public Works and Water Management for inland water in the Netherlands and a leading international center of knowledge for integrated water management.

collected (in order to be delivered to dedicated collectors) or where it eventually mixes with the bilge water. The biggest fraction (60%) of the grease, however, leaves the system on the waterside and ends up direct in the surface water. Taking into account that grease consists of 80% of mineral oil an emission factor of 58,4 kg of mineral oil per active vessel per year is estimated. This results in a total mineral oil discharge arising from the use of lubricants in the Dutch inland shipping of 58,4 x 4.571 = 266.946 kg (about 267 ton of mineral oil per year rising only from the Dutch fleet). This makes up about 9% of the total volume of mineral oil discharged in Dutch waterways from all kind of sources based on estimations for the year 2000. It is likely that this is an underestimation of the total volume of mineral oil discharges due to lubricant use by Dutch inland shipping vessels since it does only take into account the lubricants released from stern tube and rudder trunk systems. For an actual estimation additional data are required concerning accidental discharges of lubricants and lubricants released due to minor lubricating activities onboard other than the lubrication (semi) open of stern tube and rudder trunk. Note as well, that a part of the lubricants used in the stern tube and rudder trunk systems will end up in the bilge water, which is by far not always delivered to dedicated collectors. If one accounts for all possible emissions of mineral oil arising from inland shipping, including bilge water[11] and accidental spills of fuels and oily cargos (without subtracting the actually separate collected bilge water), then inland shipping would represent with 67% the biggest polluter of Dutch inland waters with mineral oil (29).

For an estimation of the total oil discharges to the European Inland Waterways due to inland shipping, data need to be collected concerning the composition and activity of the European inland shipping fleet. If we assume, however that the European Inland Shipping Fleet is in all respects comparable with the Dutch Inland Shipping Fleet than the total volume of mineral oil discharged in the European Inland Waterways due to lubricants use in inland shipping, can be derived by multiplying the total number of European active inland shipping vessel by the emission factor 58,4 kg (mineral oil /active vessel x year). At the time this report was finalized information on the total number of European inland shipping vessels was not available. Again such a calculation will be an underestimation of the total volume of mineral oil discharges due to lubricant use in inland shipping.

### 4.4.2  Sources of oil pollution for recreational boating

The findings with respect to recreational boating are mainly based on estimations done by LLINCWA in consultation with market actors and an investigation conducted by STOWA (the Dutch acronym for the Foundation for Applied Water Management Research)[12] in commission of the Waterpakt Foundation[13] (26).

---

[11]  The average annual bilgewater production per inland ship is estimated at 12.853 liter with an average volume percentage of mineral oil being estimated at 16%. Assuming an average oil density of 0,9 kg/l an emission factor of 144 kg oil per m³ bilgewater can be derived.

[12]  STOWA coordinates and commissions research on behalf of a large number of local water administrations. Among the 76 bodies that contribute to the STOWA, there are water boards, provinces and the Ministry of Transport, Public Works and Water Management.

The emissions of lubricants from recreational craft arise from two main sources: grease in the stern tubes of stern drive craft and lubricating oil emitted from two-stroke outboard engines.

### Estimation of mineral oil discharges

The Dutch fleet of recreational craft comprises ca. 252.835 boats of which 108.067 with an outboard motor and 117.640 with a stern drive engine.

Two-stroke engines are an important source of oil pollution. Two-stroke outboards not only burn their lubricating oil, but also pump much of their fuel and oil, unburned, into atmosphere and water. An estimated 30 % of all fuel and lubricating oil used in two-stroke engines ends up in the water. The contribution of two-stroke outboard engines of recreational craft to the Dutch waterways' total oil pollution is estimated to be ca. 37 ton for the year 2000 (26).

The estimation of the oil discharges arising from grease lubricated stern tubes of stern drive craft is based on the following assumptions. The first assumption is that about 40% of the stern drive craft is grease lubricated. This is a very rough (and admittedly rather uncertain) assumption based on the knowledge that stern drive boats built before 1980 are grease lubricated while boats built after this time are water lubricated. The second assumption is that an average grease lubricated stern drive recreational boat consumes annually ca. 1kg grease. 60% of the grease leaves the boat on the waterside while 40% of the grease ends up in the engine room and mixes with bilge water. At the moment, there is no regulation in the Netherlands prohibiting the pumping of bilge water into the waterways for recreational boats. It is therefore believed that the total volume of consumed grease (consisting of 80% of mineral oil) will sooner or later end up in the waterways. Accordingly, the contribution of grease lubricated stern drive recreational craft to the Dutch waterways' total oil pollution is estimated to be approximately another 37ton mineral oil per year (117.640 x 0,4 x 1 x 0,8 = 37.645 kg mineral oil per year) for the year 2000.

Consequently, the combined mineral oil emission from two-stroke outboard motors and grease lubricated stern tubes is 74ton. To this mineral oil discharges one should add the releases arising from the professional sailing ships. This is another 9,2ton mineral oil per year as it is depicted hereafter. Putting it all together the contribution of recreational boating (both privetly owned as professional) makes up about 2,8% of the total volume of mineral oil discharged in Dutch waterways from all kind of sources based on estimations for the year 2000.

---

[13]  The Waterpakt foundation is an umbrella organization of four well-known Dutch organizations for the preservation and improvement of the natural values of the aquatic environment. These organizations are: Waddenvereniging (Wadden Society), Stichting De Noordzee (North Sea Foundation), Stichting Reinwater (Clean Water Foundation) and Vereniging tot Behoud van het IJsselmeer (Society for the Conservation of the IJsselmeer).

It is possible that this is an overestimation of the actual emissions if grease lubricated crafts make up less than 40% of the stern drive lubricated recreational fleet. Moreover, not all boat owners pump their bilge water into the waterways. The last years boats are equipped with oil/water separator to draw contaminated water from bilges, capture oily contamination in a filter, and discharge only clean water. Additionally, in some yachting marinas bilge water collection systems are available for the collection and treatment of oily bilge water.

The emissions of recreational craft are very small compared with total mineral oil emissions of other main emission sources. However, this global figure can be misleading. Recreational boats are generally used in environmentally sensitive areas in pleasant weather conditions and mainly at weekends. Especially on bright summer weekends, the amount of emissions in a boating area cannot be neglected.

### 4.4.3   Professional sailing ships

The Dutch fleet that operates professionally in the water recreation market numbers about 550 vessels. Most ships are traditional sailing ships that retain their traditional character. The ships can be chartered by groups or individuals for short or long trips in the months March until October.  The majority of them operate on a grease lubricated stern tube.

With respect to lubricant use, professional sailing vessels resemble the inland shipping vessels discussed above. With respect to their activity, professional sailing vessels resemble the recreational boating sector in the sense that both professional as privately owned recreational boats navigate in the same environmentally sensitive waters.

Given an average grease consumption per professional sailing ship of 35kg per year and based on the same assumptions used in the case of inland shipping we conclude that the total mineral oil discharge arising from the use of lubricants in the sector of professional sailing ships is estimated to be ca. 9,2ton (35 x 0,6 x x0,8x 550 = 9240 kg mineral oil annually).

### 4.4.4   Conclusion

Taken individually the emissions of mineral oil from a stern tube or a two-stroke outboard engine may appear to be of negligible importance compared to an accident involving an oil tanker spill. However, when all these sources are considered as one the quantity of oil emissions proves to be of the order of 10% of the total mineral oil releases in the Dutch waterways.

4.5    BIODEGRADATION

### 4.5.1  Introduction

Biodegradability means that a substance is susceptible to biochemical breakdown by the action of micro organisms. Biodegradation represents an important route by which chemicals are removed from the environment and particularly from water. The vast amount of organisms and the complex chemical processes involved in biodegradation has caused tremendous problems and large confusion of tongues. It is the main aim in this chapter to try to untangle the main aspects of biodegradation important in the LLINCWA project.

Several types of biodegradability are distinguished. First we can consider aerobic and anaerobic biodegradation and secondly primary and ultimate biodegradation. Aerobic degradation is the breakdown of a chemical under aerobic conditions strictly meaning in the presence of oxygen, while anaerobic degradation usually means the breakdown under anaerobic conditions, meaning in the complete absence of oxygen. These conditions are specifically found in sediments and still waters. Sea sediment is a good example in this case.   Schematically primary or ultimate biodegradation is shown below.

---

Primary aerobic or anaerobic biodegradability:
Substance A $\rightarrow$ substance B
Method to analyse the disappearance of original compounds (the compound may loose its specific activity but (the rest of the molecule) may remain largely unchanged)

Ultimate aerobic or anaerobic biodegradability:
Aerobic:      Substance A $\rightarrow$ B $\rightarrow$ ...$\rightarrow$ $CO_2$ + $H_2O$ + biomass
Anaerobic:   Substance A $\rightarrow$ B $\rightarrow$ ...$\rightarrow$ $CO_2$ + $CH_4$ + $H_2O$ + biomass
Both types lead to complete mineralisation of the organic material but with different products.

---

Aerobic degradation has received by far the highest attention and will be the main focus in this chapter. Therefore we will only deal shortly with anaerobic degradation and give the main attention to the aerobic part in such a way that if we speak of primary or ultimate biodegradation we always mean the aerobic part unless specifically indicated otherwise.

### 4.5.2  Anaerobic degradation.

Anaerobic degradation occurs mainly in sediments and still deep waters where a significant part of the spilled oil will concentrate. It is difficult to measure and assess. Only recently a standard test has been developed (OECD 311 or ISO 11734). With this test the readiliy anaerobic biodegradability of chemicals is assessed. Interesting in its results is that fatty acid esters and many alcohols like 2-ethyl-hexanol are readily degraded under anaerobic conditions. This is completely different for mineral oils and polyalphaolefins. In fact their anaerobic degradation is even slower than of aerobic one (22). Combined with a high affinity of these chemicals to sediment it is most likely one of the main reason of the long persistence of mineral oil and mineral oil derivates in (sea) sediment.

### 4.5.3  Primary (aerobic) biodegradation

A first aerobic biodegradability test was introduced in 1982 for one type of oil, in response to concerns over potential environmental impacts of lubricants used in two-stroke outboard engines on the Leman Lake. This CEC L-33-T-82 test was developed especially for oils, but measures only the primary degradation i.e. loss of parent material. In this method, assessment of biodegradation of the test substance at the dose of 50 mg/litre is performed over a 21 days period using infrared spectroscopy. Test cultures are extracted with solvents (tetracloro methane or 1,1,2-trichloro trifluor ethane) and the adsorption is measured at 2930 cm −1 due to the CH3-CH2 groups present in the extract. Results are compared to poisoned controls to correct for abiotic losses of the test substance. A specific result of the CEC test on primary biodegradation is not set, but it is believed that a test result of 70% of the test compound in 21 days, and not a mixture of test compounds, is acceptable as primary biodegradable. The pass level for primary biodegradation of surfactants in detergents is proposed at 80% (2).

This method of extracting residual oil and quantification via infrared spectroscopy is widely used for monitoring hydrocarbon contamination of the water because it is easy to perform and less expensive than ultimate biodegradability tests based on for example the Sturm test method. However it measures only primary degradation and does not take into account the impact of by-products formed during biodegradation process (i.e. biodegradability or toxicity towards micro-organisms). In this way, ultimate biodegradation (see below) measurements reflect better the ability of substances to be degraded in environment.

### 4.5.4  Ultimate (aerobic) biodegradation

Ultimate biodegradation tests estimate the extent to which a substance is converted by micro organisms (the inoculum) into new biomass and simple end products of metabolism such as carbon dioxide and water. The biodegradation rate depends not only on the amount of substance (substrate) but also on the amount of micro organisms present in the test system. However, micro organisms used in these test systems also have the capacity to adapt themselves to the substance used. For example given sufficient time, specific micro organisms that use the substance to grow upon may emerge from the large variety of micro organisms originally present in the test system. Therefore different test systems are defined based on differences in concentrations of the substrate and inoculum. The OECD serial 301 A to F uses an unadapted inoculum and a low concentration of the substrate, while the OECD 302 make use of adapted inocula and/or a much higher number of micro-organisms (302A) or a higher concentration of the substance (302B and C).

In the OECD 301 series the test substance is added at a low concentration (2-100 mg/litre) to a mineral salts medium containing a mixed population of micro organisms from sludge treatment plants. The test is normally performed during a period of 28 days and biodegradability is evaluated by measurement of:
-   released carbon dioxide (301 B)
-   consumed oxygen (301 C, D and F)
-   dissolved organic carbon (301 A and E).

The low aqueous solubility of most of the base fluids used in lubricants only respirometric methods are normally used in testing, being the OECD 301 B, C, D and F or equivalent ones
The most commonly used ultimate biodegradability test is the OECD 301 B test, commonly named the modified Sturm test. It evaluates the biodegradability in aqueous media by analysis of released carbon dioxide. One can also measure the theoretical consumption of oxygen in the MITI modified test OECD 301 C, in the closed bottle test OECD 301 D and in the manometric respiration test OECD 301 F.

In the OECD 302 series the inherent biodegradability of a substance is determined (see also below). The test conditions do not simulate those conditions experienced in a sewage treatment plant. Test conditions are stimulating the adaptation of the microorganisms to the substrate by either using a high concentration of the inoculum and/or a high concentration of the substrate.

### 4.5.5   Ready (bio)degradability

There is a legal requirement within the European Community to assess the biodegradability of chemicals marketed at more than 1 tonne /year, using some form of aquatic screening test which provides a measure of ultimate biodegradation. The biodegradability of the EU hazard classification systems is based only on the distinction between the ready biodegradability of a substance or not. It can therefore only be applied to a single substance. In other words, a biodegradability test can be performed on the total product, but interpretation in terms of a EU classification can only be based on the single substance.
Nowadays ready biodegradability is measured using one of the OECD 301 or equivalent tests. If the substance in such a test is degraded within 28 days for 60% by the OECD 301 B, C, D and F or 70% by the OECD 301 A and E, and this level was attained within 10 days after 10% of the substance has disappeared, the substance is called readily biodegradable. The last requirement is normally called the 10day-

window and indicates the biodegradation rate. However, this requirement will not be necessary for a number of surfactants in detergents in EU regulation (2).

In the past tests were introduced to measure the biological oxygen demand (BOD) of the substance in a specific number of days like the ISO 5815. Therefore the term readily biodegradability can also be defined as the ratio of the BOD in 5 days (BOD5) and the chemical oxygen demand (COD), measured by e.g. the ISO 6060 test. If this ratio is larger than 0.5, the substance is also called readily biodegradable.

In fact other degradation mechanisms are also known like hydrolysis and photolysis. Therefore if standard tests indicate that the substance is mineralised by any other way within 28 days for more than 70%, it is also called readily degradable.

It is generally assumed that the substance in the water compartment will be degraded sufficiently fast in order not to cause long-term adverse ecotoxicological effects. However, it does not mean that any chemical, which is not readily biodegradable will not biodegrade at all. In fact nearly all still do, but on a much lower scale. They are then generally "inherent" biodegradable.

### 4.5.6 Inherent biodegradability

Since inherent biodegradability can be considered to be a specific property of a chemical, it is not necessary to define limits on test duration or biodegradation rate at this level of testing. A figure of more than 20% biodegradation in these tests may be regarded as evidence for inherent, primary biodegradability. A figure of more than 70% mineralisation may be regarded as evidence for ultimate biodegradation. The Zahn-Wellens/EPMA test (302 B) defines the validity of the test (inherently biodegradable) at 70% removal within 14 days and a gradual removal of DOC (dissolved organic carbon) over days and weeks. Both, the 302 A and B tests, use the removal of DOC to assess the results. For test compounds with a low solubility like mineral and vegetable oils such test methods are not suitable. This leaves only the 302 C as possible alternative but this one is reported to have fallen into disuse. A modified version of ISO 14593 is recommended by CONCAWE for assessing the inherent biodegradability of lubricants. Pass rates of 60% of the theoretical maximum of carbon dioxide within 28 days is taken here as evidence of ultimate inherent biodegradability while less than 20% indicates that the test compound is not inherent biodegradable under the test conditions.

Such results are of importance in risk assessments, because given a specific time a substantial amount of the substance will still be degraded, diminishing possible adverse long term effects on organisms. However, in the EU hazard classification inherent biodegradable results are not included. Extrapolation of the results of inherent tests to risk assessments should be done with great caution because of the strongly favourable conditions for biodegradation that are present within these tests.

### 4.5.7 Comparison of biodegradability values of different base fluids

In lubricants mainly mineral oil base fluids are used. However new types of base fluids are introduced with a similar or better technical and environmental performance. They are mentioned in Chapter 2. A comparison of primary and ultimate biodegradation test results is shown in Table 1

Table 4.1.    aerobic and anaerobic biodegradability values of mineral oils and vegetable based oils (3)

| % Biodegradability | Mineral oil | Polyalpha-olefin | Polyinternal olefin | Oleochem-ical esters | Vegetable oils |
|---|---|---|---|---|---|
| primary (CEC L 33-A-93) | 20-30 | 25-35 | 20-60 | 85-95 | > 95 |
| Ultimate aerobic (OECD 301 B) | 20-35 | 30-70 | 30-65 | 85-95 | > 95 |
| Ultimate anaerobic (ECETOC)[22] | 4-6 | 0-14 | 20-50 | unknown | 80-90 |

Table 1 shows a good correlation between CEC and OECD tests for the presented base fluids. However such a correlation is not present in other cases (4). A well-known example is the high primary biodegradation of most alkylphenol ethoxylates (~90%) and the low ultimate biodegradation (< 50%). Both methods show high biodegradability values for **esters and vegetable oils**. On the contrary the complex mixtures of hydrocarbons present in mineral oils are only partially degraded. These results can be explained fairly well from the current knowledge of microbial metabolism.

The first step in the degradation of vegetable based oils involves enzyme-catalysed cleavage of the ester bond to fatty acids. The high biodegradability of vegetable oils or based esters, is to be expected, given that involved enzymes, esterases and lipases, are produced by a wide range of micro-organisms. Then both saturated and unsaturated fatty acids are readily mineralised via a process of β-oxidation.

In mineral oils, n-paraffins degradation occurs in the same β-oxidation pathway. The first step is, however, slow owing to the lack of an ester group in the molecule and alicyclic rings are relatively resistant to microbial attack owing to structural hindrance.

## 4.6    ECOTOXICITY

### 4.6.1  Introduction

In toxicological testing one generally tries to determine dose-effect relationships. This can be done for terrestrial animals to which the substance tested can be administered orally or by injection. In aquatic ecotoxicological testing this is virtually impossible, particularly for small animals, because the test substance will be administered with water; concentration-effects are then determined. Although dose-effect relationships in aquatic ecotoxicology should be studied by determining analytically the amount of the test compound in the test organisms, this is not generally done.

In extrapolating the results of ecotoxicological tests, the fact that the test compound may have been changed in the environment should be taken into account. Therefore the concept "dose" will indicate amounts administered directly to or present in the test organism, while the amounts present in the test medium or in the environment is called "environmental concentration". Physical-chemical properties like volatility, hydrolysis, oxidation etc) and environmental factors like adsorption and biodegradability will determine what fraction of the initial environmental concentration will be active in the immediate vicinity of the living system. In standard tests these effects should be taken into account or eliminated as much as possible.

In general biodegradability of a test compound is checked before its testing on aquatic toxicity. Biodegradability may deplete the amount of oxygen in the water and the tested organism may be affected by this depletion. Clearly interpretation of the test result in that case leads to incorrect ecotoxicological values.

In ecotoxicity the following tests are common:
* short term tests with single species
* long term tests with single species
* tests with multiple species
* tests with infra-organisms (e.g. cell or tissue cultures).

Special care should be taken to ensure that the environmental concentration of the test compound remains constant during test periods. Generally, for aquatic organisms, the following distinction is made:

* static tests (the test compound is added once to the test system, no flow occurs and the test medium is not changed
* semi-static tests in which the test-medium and the compound are periodically replaced
* flow-through (continuous supply of a constant concentration of the test compound in the test system to the organisms exposed.

### 4.6.2  Ecotoxicity tests

In ecotoxicology the most difficult situation is that which involves the prediction of the impact of very low concentrations of a chemical, maintained for long periods, and affecting communities of organisms. Such standard tests are simply not available owing to its complexity. Because of such factors it is generally agreed that a stepwise procedure is to be followed, starting usually with simple short term tests for all compounds with single species but terminating with more complex multi-species and even field tests for those compounds which are of great practical value but which have been indicated as being potentially dangerous to the environment.

A stepwise procedure might involve three steps: the basic level, the confirmatory level and the definitive level. The following basic level tests are proposed and included in the EU hazard assessment of substances: Alga growth inhibition test (OECD 201), Daphnia acute immobilisation test (including reproduction test) (OECD 202) and fish acute toxicity test (OECD 203). The reason for using more than one test species is that the variation between species in sensitivity to one type of toxic chemical need not necessarily coincide with the variation in sensitivity to other types of chemical. It is also not practical to prescribe only one test species to be used all over the world. However, the above three species are chosen within the EU for the water compartment. These basic level tests all measure an acute effect of the test compound on the organism within a relatively short period. In the EU requirements the 50% effect level is chosen; the level at which 50% of the test organisms show an adverse (lethal) effect. The test procedure prescribes a specific test duration, but this varies depending on the specific aims. The EU prescribes for the OECD 201 a test period of 72 hours, for the OECD 202 A of 48 hours and for the OECD 203 96 hours.

At the confirmatory level tests should be used which yield more complete information if suspicion as to the acceptability of a chemical has been previously raised at this basic level.

Testing at the definite level may be needed in some special cases, e.g. where appreciable environmental concentrations of the chemical are likely to be involved and/or some indication of possible environmental hazards exists.

The following list presents an overall view of the OECD ecotoxicity standard tests:

**A/ eco-toxicity tests on micro-organisms:**
- OECD guideline 201 : Alga, Growth Inhibition Test.
- OECD guideline 202 : Daphnia sp. Acute Immobilisation Test and Reproduction Test.

These two tests are examples of short-term tests with single species and a single chemical.

- OECD guideline 216 : Soil Micro organisms, Nitrogen Mineralisation Test.
- OECD guideline 217 : Soil Micro organisms, Carbon Mineralisation Test
- OECD guideline 209 Activated sludge, Respiration inhibition test

These three tests are examples of short-term tests with multiple species and a single chemical.

- International Standard ISO 11348 : Water quality – Determination of the inhibitory effect of water samples on the light emission of Vibrio Fischeri (luminescent bacteria test).

This is an example of a short-term test with a single substance and multiple chemicals.

**B/ eco-toxicity tests on superior organisms :**
- OECD guideline 203 : Fish, Acute Toxicity Test.
- OECD guideline 207 : Earthworm, Acute Toxicity Test.

They are short term tests with a single chemical and organism

- OECD guideline 215 : Fish, Juvenile Growth Test.

A long term tests with a single chemical and organism

- International Standard ISO 7346-2 : Water quality : determination of the acute toxicity on fish Brachydanio rerio Hamilton-Buchanan (Téléostei, Cyprinidae). Semi-static method.

A short term test with a single specie and multiple chemicals

### 4.6.3   Substances with a low water solubility

The experimental investigation and the assessment of the ecological properties of substances with a (very) low solubility (like oleo chemical esters or mineral oil ) are not trivial.

The determination of $EC_{50}$-values *(efficient concentration)* for algae and daphnia can be biased by insoluble portions of the test substance floating on the surface of the test media. An overestimation of the toxicity may be the result. In this case, a special test protocol for water insoluble substance exists. It is based on testing of water accommodated fraction (WAF). WAFs are prepared by stirring the test substance for at least 24 hrs in water and subsequently removing the insoluble portions by an appropriate method. It has been shown that WAFs up to a loading of 100 mg/l are not toxic, an $EC_{50}> 100$ mg/l can be used for environmental classification.

### 4.6.4   Comparison of acute aquatic toxicity values of different base fluids

Mineral oil is a complex mixture of aliphatic, aromatic and cyclic hydrocarbons and its composition varies substantially depending on its application. Lubricating oils and greases are mainly composed of high boiling hydrocarbons (80% > $C_{22}$ and 20% < $C_{22}$). After emission in the environment its composition is also changing continuously owing to physical-chemical effects. Owing to its complex composition, MTR, MTC or NOEC-values of mineral oil are extremely difficult to determine. The toxic effects of mineral oil can be divided into three main aspects: *a)* a specific mode of action on specific biochemical processes in organisms, *b*) an a-specific mode of action: general narcotic effects owing to the accumulation of substances in the cell membranes (generally known as basic or minimum toxicity or narcosis) and *c)* physical effects owing to the staining of organs caused by the viscous character of mineral oil. Effect *a* and *b* can be explained by the presence of specific components in the mineral oils. Especially the very light carbon-fraction (C10-C14) and specific mono- and di-aromatic compounds are most likely responsible. Standard ecotoxicity tests point into this direction.

The presence of a separate oil phase (as droplets or as coating on sediment particles) seems to involve the staining effects. Owing to contact activities essential functions of organisms living in the aquatic or sediment may be lost partially or fully like respiration, locomotion or food intake. These adverse effects are presently not considered in ecotoxicological tests. However, microscopic research in the physical appearance of mineral oils in field sediments indicate that 70% consists of droplets. This percentage seems to be directly proportional to its concentration. Similarly the size of the droplets also increases on an increase in concentration (5). It is the staining effect that mainly causes the adverse ecotoxicological effects of mineral oils used as lubricant oils or greases.

Vegetable oils do not pose these problems. They are biodegraded relatively fast and are used as food source for organisms.

Within the LLINCWA project ecotoxicity values have been determined on fresh and used samples of biolubricants. Tests on the marketed substances are required for classification and labelling. The possible change of ecotoxicological values due to its use is not incorporated in the classification, although it is often these residues that are lost into the environment.

Table 4.2 shows the results of these tests practised on some base fluids used in biolubricants:

Table 4.2: Ecotoxicological properties of representative oleo chemical esters (6)

|  | Substances | Fish toxicity ($LC_{50}$) mg/l | Daphnia toxicity ($EC_{50}$) mg/L | Bacteria toxicity ($EC_{50}$) mg/l |
|---|---|---|---|---|
| Fatty acid ester | 2-ethyl hexylcocoate | 10 000 | >>water solubility * | 10 000 |
| Glycerol esters | Glycerol trioleate | 10 000 | >>water solubility * | 10 000 |
| Polyols esters | Trimethylol propane-trioleate TMP mixed | 5 500 | > 1 000 mg/l | 10 000 |
| Complex esters | Esters with adipinic and oleic acid | 5 500 | >>water solubility * | 10 000 |

* water accommodated fraction (WAF) tested after removal of insoluble organic material

Although the algae test results are not indicated on Table 4.2, the results show values much higher than the 100 mg/L limit necessary for a possible classification. The above base fluids therefore do not contribute to the ecotoxicity of the biolubricant. However, testing of the original lubricant may indicate a higher ecotoxicity and it is expected that in those cases  added chemicals like the thickener and/or additives will have caused this increase. Such a picture emerges in the following table where the original and used lubricants are tested on their ecotoxicity.

Table 4.3 shows behaviour of 3 biolubricants before and after operation 1000 hours as hydraulic fluids on forestry machines:

Table 4.3: wear effect of biolubricants on ecotoxicity properties (6)

| Lubricant | | Algae toxicity $(EC_{50}$-72h) OECD 201 | Fish toxicity $(LC_{50}$-48h) OECD 203 | Daphnia toxicity $(EC_{50}$-48h) OECD 202 |
|---|---|---|---|---|
| Biohydran | new | 2800 | >10 000 | >10 000 |
| | used | 1800 | >10 000 | >10 000 |
| Biolube | new | 5400 | >10 000 | >10 000 |
| | used | 5600 | >10 000 | 5 900 |
| Helianathe | new | 4800 | >9 900 | >9 900 |
| | used | 100 | >10 000 | 2 500 |

From Table 4.3 it can be concluded that the Algae toxicity test seems to be the most sensitive one of the three tests for the three tested lubricants. Such a general picture is also confirmed by a larger study on ecotoxicity data of HPV-chemicals present in the EU IUCLID database (7). The table also shows that the ecotoxicity is increasing while using. The cause of this increase is unknown since many different chemical processes may occur during wear. However, the increase in ecotoxicity on use does not fall below the limit of 100 mg/L.

The ultimate toxicity of the mixture (biolubricant) is not the sum of the individual toxicities (additives), because phenomena like bio availability, synergistic or antagonistic effects dictate the ultimate hazard of the mixture. The only way to get a real answer to the environmental hazard of multi-chemical products is the achievement of ecotoxicity bioassays. However in the EU the hazard classification is obtained by assuming an addition of ecotoxic effects of the individual components.


4.7    CLASSIFICATION SYSTEMS


4.7.1    **European developments**

In the last four years three aspects can be identified as important in relation to lubricants:
- The publication of the new directive on labelling products and preparations including biolubricants.
- The establishment of a priority list for substances for the aquatic policy.
- The acceptance of the product group *lubricants* for the EU- ecolabelling process (December 2002)

Different EU directives are dealing with the hazards of substances and preparations. A recent and important change was published with the EU Directive 1999/45/EC[14], the European preparation directive. In the new directive environmental hazards for the aquatic compartment and ozone layer are taken into account and new guidelines on the availability of Material Safety Data Sheets (MSDS) are formulated. The estimation of

---

[14] Directive 1999/45/EC of the European Parliament and the Council, 30.7.1999

the environmental hazards as well as the health hazards is being based on the individual ingredients and their concentration in the product (in this case in the biolubricant). The assessment of the aquatic hazards is mainly based on data of the acute aquatic toxicity for fish, Daphnia's and algae, ready biodegradability and bioaccumulation. New risk phrases are developed that describe the specific (environmental) risks. Hazard assessment  may lead to a classification of the biolubricant according to the risk phrases R50, R51, R52, R53 or a combination of these risk phrases.

A practical problem that arises is that there are many substances that lack enough environmental data. Especially "older" substances, those that are in the market longer than 1981 may suffer this problem and as a consequence these substances will not have been assigned with any risk phrase that therefore will not be described in the MSDS. For new substances, introduced in the market after 1981, no data will lead to a warning phrase on the label. Therefore due to a possible lack of data for old substances, the information on a MSDS is no guarantee for a harmless substance and might lead to an underserved *unharmless* estimation.

The publication of the new EU Directive has a strong impact on the existing biolubricants ecostandards. The Nordic Swan and the Swedish Standards are brought into line with the new directive. However they demand the availability of all requested data for all components of the lubricant, also for the older substances.

The establishment of a priority list for substances for the aquatic policy is important for the formulation of lubricants[15]. Based on its intrinsic properties, its occurrence in the aquatic compartments and other existing regulations a list of 32 substances was identified whose emission into the water compartment should be phased out.

The EU new preparation directive and the EU list of priority substances are basic to the development of the LLINCWA classification system.

### 4.7.2   Presently existing ecostandards

Several countries have developed ecostandards for biolubricants. Most of them have a similar set up, but the developed criteria show large differences. All take into account biodegradation, acute aquatic toxicity, health hazards, renewable resources and the technical performance. The main differences are caused by the interpretation of some environmentally relevant parameters. A common problem is the lack of needed data on used ingredients in biolubricants.   In Europe the following ecostandards or ecodirectives on biolubricants are present (December 2002); the Nordic Swan, the Swedish standard for hydraulic fluids and lubricating grease, the German Blue Angel and the Dutch Vamil. Additionally, in many countries an ecolabel for chainsaw oils has been developed (for France this was done very recently (December 2002)), but these are not summarised in the following table because this application does not fall within the scope of the LLINCWA project.

The main characteristics of the existing ecostandards are shown below.

---

[15] Brussels, 07.02.2000, COM(2000) 47 final,  2000/0035 (COD)

| Ecolabel | Application | Assess-mentBase | Biode-gradation | Aquatic toxicity | Health hazards | Renewa-bility | Technical perfor-mance |
|---|---|---|---|---|---|---|---|
| Nordic Swan | Hydraulic fluid Gear oil Lubricating grease 2-stroke oil | Ingredients | X | X | X | X | X |
| Swedish standard, SS 155434 | Hydraulic fluid | Ingredients | X | X | X | | X |
| Swedish Standard, SS 155470 or Gothenburg system | Lubricating grease | Ingredients | X | X | X | X | X |
| German Blauer Engel | Hydraulic fluid Lubricating grease | Ingredients | X | X | X | | X |
| Dutch VAMIL | Hydraulic fluid Lubricating grease | Product | X | X | | | |

The Swedish standard SS 155434, the German Blue Angel and the Dutch VAMIL standards are successful ecostandards. Quite a number of biolubricants comply with these criteria. Their criteria have been incorporated in the LLINCWA classification system. This system therefore follows the actual ecolabels of the different European member states and as a consequence this means that those biolubricants that already posses one of the mentioned ecolabels are automatically assigned to a class in the LLINCWA system.

However, the large differences in used criteria do emphasize the need for a uniform harmonised classification system, leading to a uniform use of the term biolubricant. A European ecolabel could be an answer to this urgent need.

### 4.7.3    LLINCWA Classification system

Criteria are set for different types of biolubricants, since their composition does vary substantially. In the LLINCWA project criteria have been set for both bio-greases and bio-gear oils, and bio-hydraulic fluids, lubricants that are use in or near the inland or coastal waters.

For bio-hydraulic fluids four categories are defined varying from BH1, the most desired bio-hydraulic fluid, to BH4, still acceptable as bio-hydraulic fluid. For greases and gear oils three classes are defined, varying respectively from BG1 to BG3.

Starting point is that in biolubricants no components may have been added that are found on the EU list of priority substances in the field of water policy[16].

---

[16] See: http://europa.eu.int/

The criteria are compared to existing European (eco)labels or -standards on biolubricants. Some of these bear a sufficiently great resemblance in order to assign products with these eco-criteria automatically to one of the defined classes. In case a biolubricant does not belong to one of these classes, it should fulfil the required criteria. It can be concluded that for most classes in the LLINCWA classification system a number of biolubricants are already on the market.

The highest categories are defined based on a criteria set that determines properties for individual components. This means that, in accordance with the new preparation directive, for all individual components environmental and health data should be available. In the lowest biolubricant classes criteria for the whole product are defined. In this case it is assumed that the biodegradation and toxicity tests are performed on the total product.

The column *"performance"* is actually not discriminating between *bio*lubricants and *non-bio*lubricants. It is added on request of suppliers to show that the "ranked" biolubricants comply with demands of the original equipment manufacturers or are according to internationally accepted standards.

# Classification system for bio-hydraulic fluids

| Category | Comp./Prod.* | Re-newable** | Biodegradation | Aquatic toxicity | Health and other hazards | Performance | Similarity with ecolabel |
|---|---|---|---|---|---|---|---|
| BH-I | C | 65% | | The total amount of components with a R-phrase of R50*** or a combination thereof should not exceed 1%, of R51/53 not more than 2% and of R53 not more than 3%. (w/w) | No components with R-phrases or combinations thereof in relation to sensitization, carcinogenity, mutagenicity and reprotox (R39, R40, R42, R43, R45, R46, R48, R49, R60-R64). Other components comply with EU directive on preparations. No R –phrase for the end product. | SS 155434 or VDMA 24568 or supplier/OEM warrant guarantee | Swan |
| BH-II | C | - | | The total amount of components with a R-phrase of R50*** or a combination there of should not exceed 1%, of R51/53 not exceeding 2% and of R52 and/or R53 not exceeding 3%. (w/w) | No components with R-phrases or combinations thereof in relation to sensitization, carcinogenity, mutagenicity and reprotox (R39, R40, R42, R43, R45, R46, R48, R49, R60-R64). Other components comply with EU directive on preparations. No R –phrase for the product. | SS 155434 or VDMA 24568 or supplier/OEM warrant guarantee | SS 155434 |
| BH-III | C | - | Primary (>80%) degradation: total >95% | $XC_{50}< 1$mg/L max 1%   1 mg/L $\leq XC_{50} \leq$ 100 mg/L: max. 5%   $XC_{50}>$ 100 mg/L $\geq$ 95% (OECD 201, 202 and 203) | No components with R-phrases or combinations thereof in relation to sensitization, carcinogenity, mutagenicity and reprotox (R39, R40, R42, R43, R45, R46, R48, R49, R60-R64). Other components comply with EU directive on preparations. No R –phrase for the product except R65 | SS 155434 or VDMA 24568 or supplier/OEM warrant guarantee | Blauer Engel |
| BH-IV | P | - | Primary (>90%) degradation. | $XC_{50}>$100 mg/L according to OECD 201 and 202 | No components with R-phrases or combinations thereof in relation to sensitization, carcinogenity, mutagenicity and reprotox (R39, R40, R42, R43, R 45, R46, R48, R49, R60-R64) | SS 155434 or VDMA 24568 or supplier/OEM warrant guarantee | VAMIL |

\*   C means: criteria are set for all individual components in the product. P means criteria are set for the total product (irrespective of the used components)

\*\*   Only for the first category it is obligatory that the biolubricant is base > 65% renewable raw materials (Vegetable oils).

\*\*\*   The assignment of an environmental hazard R-phrase R50 implies ready biodegradability.

- Ready biodegradable according to OECD 301 or equivalent tests like BODIS (ISO 10708), ISO 9408 and ISO 9439.
- Primary biodegradation tests should lead to the indicated percentage within **28** days.
- Components should not be found on the EU list of priority substances in the field of water policy.
- No components means less than what is stipulated by the current EU preparation directive

# Classification system for bio- gear, lubricating oils and bio-greases

| Category | Comp./Prod. * | Re-newable ** | Environmental hazards | Health and other Hazards | Performance | Similarity with ecolabel |
|---|---|---|---|---|---|---|
| BG-I | C | 65% | No components with a R-phrase or combinations thereof in relation to very toxic for the water compartment: N, R50***. Other components comply with EU directive on preparation. No R –phrase for the end product. | No components with R-phrases or combinations thereof in relation to sensitization, carcinogenity, mutagenicity and reprotox (R39, R40, R42, R43, R45, R46, R48, R49, R60-R64). Other components comply with EU directives on preparation. No R–phrase for the end product. | Warrant guarantee is given by supplier/OEM | Gothenburg B or SS15 54 70 (Class B) |
| BG-II | P | Base fluid from vegetable oil or synthetic esters. | Biodegradation: Ultimate: OECD 301B,C,D,F >70% or OECD A,E>80% or primary: CEC>90% or equivalent tests — Aquatic toxicity XC$_{50}$>100 mg/L for OECD 201 and 202 or WGK$\leq$1. | No components with R-phrases or combinations thereof in relation to sensitization, carcinogenity, mutagenicity and reprotox (R39, R40, R42, R43, R45, R46, R48, R49, R60-R64). Other components comply with EU directive on preparations. No R-phrase for the end product except R10, R65, R66 | Warrant guarantee is given by supplier/OEM | Blauer Engel. |
| BG-III | P | - | Biodegradation: Ultimate: OECD 301B,C,D,F >60 or OECD A,E>70% or primary: CEC>90% or equivalent tests — Aquatic toxicity XC$_{50}$> 1mg/L for OECD 201 and 202 | No components with R-phrases or combinations thereof in relation to sensitization, carcinogenity, mutagenicity and reprotox (R39, R40, R42, R43, R 45, R46, R48, R49, R60-R64) | warrant guarantee is given by supplier/OEM | VAMIL |

\*   C means: criteria are set for all individual components in the product. P means criteria are set for the total product (irrespective of the used components)

     Only for the first category it is obligatory that the biolubricant is base > 65% renewable raw materials (Vegetable oils).

\*\*   The assignment of an environmental hazard R-phrase includes ready biodegradability aspects.

\*\*\*

- Ready biodegradable according to OECD 301 or equivalent tests like BODIS (ISO 10708), ISO 9408 and ISO 9439.
- Primary biodegradation tests should lead to the indicated percentage within **28 days**.
- Components should not be found on the EU list of priority substances in the field of water policy.
- No components means less than what is stipulated by the current EU preparation directive.
- A list of products may be obtained from the following website:

    **BG1:** Gothenburg http://www.gbgreg.kommunalforbund.se/regionalplanering/miljo/miljofetteng.html

    BG2: The Blauer Engel  http://www.blauer-engel.de/English/index.htm.

    BG3. The Dutch VAMIL http://www.vamil.nl.

### 4.7.4   The German Wassergefährdungsklasse

A national German initiative to classify aquatic hazards of a substance in water is the Wassergefährdungsklasse (WGK), since 1999 harmonised with existing EU criteria. It defines a water hazard class for a substance based on a combination of aquatic and human toxicological and (physical)-chemical properties. These properties correspond with a subset of the EU R-phrases. However, lack of specific data of the substance may lead to an increased hazard classification.

The classification is committed according to the Administrative Regulation on the Classification of Substances Hazardous to Waters into Water Hazard Classes (Verwaltungsvorschrift wassergefährdende Stoffe - VwVwS) of 17 May 1999. With the VwVwS that has been entered into force on 01 June 1999 the former WGK 0 (not generally hazardous to waters) is not further continued. Instead a classification as "non-hazardous to waters" ("nicht wassergefährdend" - nwg) is introduced.

The German Ministry of Environment does administer the WGK classification system and takes care for a regularly updating. Three classes are defined in the new WGK system.

- WGK 1: Low water hazardous
- WGK 2: Water hazardous
- WGK 3: Severly water hazardous

A similar classification system has been set recently in the Netherlands as well to assess substances and products within the framework of emission permissions for water pollution (11).

### 4.7.5   ISO Standard

The International Standard specifies the requirements for environmentally acceptable hydraulic fluids and is intended for hydraulic systems, particularly hydrostatic (24). This standard stipulates the requirements for environmentally acceptable hydraulic fluids at the time of delivery, the ready for use product.

The ISO standard presents specifications for the following types of hydraulic fluids:
HETG hydraulic fluids, non-water-soluble, vegetable oil types.
HEES hydraulic fluids, non-water-soluble, synthetic ester types
HEPG hydraulic fluids, water-soluble, poly glycol types.
HEPR hydraulic fluids, poly alpha olefins and related hydrocarbon products.

| Characteristics of Test | Units | Requirements | Test method or standards |
|---|---|---|---|
| Biodegradability | % | 60 | ISO 14593 or ISO 9439 |
| Toxicity[a]<br>Acute fish toxicity, 96h, LC50, min.<br>Acute Daphnia toxicity, 48h, EC50, min. | <br>mg/l<br>mg/l | <br>100<br>100 | <br>ISO 7346-2<br>ISO 6341 |
| Bacterial inhibition 3h, EC50 min. | mg/l | 100 | ISO 8192 |
| [a] Water-soluble fluids shall be tested according to the test method cited. Fluids with low water solubility shall be tested using water-accommodated fractions, prepared according to ASTM D6081 | | | |

Table, 4.4-    Environmental behaviour requirements for categories HETG, HEES, HEPG and HEPR

ISO 14539, the biodegradability test stipulated in the ISO 15380 standard, is a ready biodegradability tests; a recent variation of OECD 301D. It uses closed flasks and measurement of the inorganic carbon (IC) released by biodegradation in the headspace (flasks are sacrificed at each measurement) and uses no adaptation in advance of the test and no 10-day window.

ISO 7346-2 and ISO 6341 - the toxicity tests postulated in ISO 15380 – are both acute toxicity tests. ISO 7346-2 is an acute lethal toxicity test of substances to a freshwater fish [Brachydanio rerio Hamilton-Buchanan (Teleostei, Cyprinidae)] while ISO 6341 determines the inhibition of the mobility of Daphnia magna Straus.

All above-mentioned protocols/tests are designed to assess the biodegradability and toxicity of single substances. In ISO 15380 these protocols are applied to the formulated lubricant. Assuming that the base oil mainly determines the overall biodegradability of a lubricant, testing the formulated lubricant (mixture of base oil and additives) according to ISO 14593 is a fair way to assess its overall biodegradability. But is it as fair to test the end product for toxicity if we know that it are the additives that mainly determine the toxicity of a formulated lubricant? According to Battersby (4) this is highly depended on the selected technique for dosing during the performance of the tests. The preferred technique for dosing when applying the above mentioned protocols/tests to the formulated lubricant is to present the lubricant as water accommodated fractions (WAFs). As general rule, for formulated lubricants, the predominant material in the WAF is likely to be the additives and it is this material that may cause any toxicity measured.

Thus, a legitimate argument not to use the biodegradability and toxicity assessment of lubricants according the ISO 15380 is not that this approach may result in an underestimation of the actual environmental burden caused by the formulated product as far as the lubricant is presented as WAF. The argument not to use the assessment procedure set forth in ISO 15380 is that additional information on the environmental performance of the separate ingredients is required for the purposes of the classification, packaging and labelling of dangerous preparations according to the revised dangerous preparations directive (1999/45/EC) as well as according to some eco-labelling schemes.

Additionally to the environmental and technical standards the ISO-standard presents a guideline for changing for changing fluids from mineral-based oil to environmentally acceptable fluids; a useful tool for users that actually make the choice to substitute. A comparable tool is presented in chapter 8.

### 4.7.6  The European White Paper

An important drawback on the EU classification system, but in fact for all presented classification systems,  is the consequence of the division between old and new chemicals. Old chemicals are substances that were already on the market before 1981, new chemicals are those that are introduced at the market after that time. Presently around 100 000 substances belong to the category *old substances*. For this group it was allowed to limit the risk or hazard assessment of these substances only to the available data. As a consequence it signifies that in case no data are available no risk phrase can be given and this is (until now) acceptable for the authorities.

Although a lot of toxicological and environmental data became available after 1981, still many are missing. For example for the High Production Volume (HPV) chemicals only 46% of the needed data on the algae toxicity is available, 55% of the data on acute toxicity to aquatic invertebrates (like daphnia's) and 68% on the acute/prolonged toxicity of fish (9).

Of the (for risk assessment) requested three standard tests it has also been shown that the algae toxicity value is the most sensitive, meaning that omission of this value will lead to the largest variation in change of environmental hazard classification. It is expected that the availability of the necessary environmental data will be even substantially lower for non-HPV-chemicals.

New substances in the EU are those that entered the market after 1981. Nowadays around 27.000 substances belong to this group. For those chemicals much more data need to be handed over to the EU authorities before it is allowed on the market. These data are used for the establishment of a risk assessment of the chemical. If insufficient data would be available to determine its hazards and if the substance is present in the product at a concentration level higher than 1%, it is obliged to indicate this on the label of the preparation. Than the label must bear the inscription: *"warning- this preparation contains a substance not tested completely"*.

It is, among others, this discrepancy and the previously mentioned point that only around 20% of the analysed chemicals can explain the biological ecotoxicity (in water) that stimulated the creation of ecolabels, national initiatives like the establishment of the German WGK system and the EU White paper strategy for a future chemicals policy (10). The great difference is that if the necessary data are lacking or of insufficient quality, the product or substance cannot be given an ecolabel or is being much higher ranked in the established classification system. For example, in the case of the German WGK system, a substance with lacking data will be ranked as WGK3, the highest possible hazard value.

The use of a specific subset of test data or R-phrases to estimate the hazard or risk of a substance for a specific compartment or in total is also highlighted in several other initiatives. The EU white paper for a future strategy of chemicals policy is intending to establish a risk, evaluation and authorisation of chemicals REACH system for all present substances on the EU market, based on data as defined in ANNEX VII A of 67/548/EEC. The system is currently debated.

In the Netherlands a hazard concern classification system has been developed recently as part of a new environmental policy, called strategy on managing substances (SOMS). This system is using a subset of health hazard and environmental hazard data (5).

## 4.8    CONCLUSIONS ENVIRONMENTAL ASPECTS

Mineral oil and synthetic hydrocarbons are important water polluting substances in European waters. Educated guesses estimate a total yearly use of 5 million tons of lubricants in all applications: land and water. Suppliers do not distinguish between land and aquatic use, which makes it impossible to make a clear estimation of the yearly lubricants use in the aquatic area. Around 45% of these lubricants is being spilled somewhere in the environment: deliberately or accidentally, showing the enormous need to stimulate the use of biodegradable, non-toxic lubricants.

Measurements show a total load on mineral oil of 1032 tons/year (1997) entering the Netherlands as pollution in Rhine water, on a total pollutant load of more than 5.000 tons/year. This means a load of more than 20% due to mineral oil originating from all sources machine oils, spills from metal working fluids, lubricants etc. Added to this are emissions in the Netherlands itself. A calculation of the actual spill caused by inland shipping in the Netherlands shows a significant contribution to the total mineral oil load. (Loss) lubrication of the screw axe and rudder system of the Dutch inland fleet does contribute annually around 300 ton to the actual aquatic pollution with mineral oils. In the past the total oil spill of the inland fleet was considerably larger, due to direct spill of the bilge water as well. With the separate collection of bilge water a strong reduction is realised, although still a considerable amount of bilge water is spilled in inland water (in 2000 more than 300 bilge water spills were registered in the Dutch inland waters). The recreational sailing activities in the Netherlands, with their screw axes and still a high use of 2-stroke outboard engines, contribute a significant amount of 83 tons mineral oils, mainly emitted in the more environmentally sensitive areas.

Additional to these sources are the mentioned structural and accidental sources from inland water management, hydroelectric power generation and of course many on-land activities with a discharge to the fresh waters. Unfortunately it was not possible to estimate their separate to the aquatic pollution.

Compared to base fluids used in biolubricants the spilled amount of mineral oils is disproportionately larger while they show a low aerobic and anaerobic biodegradation. Although the exact contribution of mineral oil based lubricants to the total aquatic mineral oil pollution is not easily to assess, it can be concluded that they make up a substantial part of the diffuse water pollution. However, it is not only mineral oil based base fluids in lubricants that determines the environmental hazard of lubricants, several additives especially some used in hydraulic fluids, give a substantial contribution to the toxicity of the spill.

The main environmental problems with mineral oil used in hydraulic fluids, gear oils, 2-stroke oils and greases are highlighted in its physical effect of staining essential organs and its low biodegradation (both aerobic and anaerobic). According to the EU criteria its acute aquatic toxicity is too low to classify many mineral oil distillates as hazardous for the (aquatic) environment.

Vegetable oils and synthetic esters have a much better biodegradation capacity than mineral oil under aerobic as well as anaerobic conditions. They are usually classified as readily biodegradable and due to this property staining effects will have a much lower impact on aquatic organisms as well. Their aquatic toxicity is comparable or lower than of mineral oil.

A substitution of mineral oil by base fluids that are readily biodegradable and whose aquatic toxicity is similar or even lower than of mineral oil does increase substantially the environmental performance of the lubricant. They have the advantage that long-term effects due to the base fluid can be ruled out.

The used additives in the lubricants, mineral as well as bio, therefore largely determine the actual toxicity of lubricants. Although this does attach the area of confidentiality, a systematic screening of used additives is strongly advisable to be able to choose for low toxic and biodegradable additives, to comply with environmentally classification systems. A proper selection or specific development of base fluids for biolubricants may lead to a reduced need to apply additives. In environmental and in toxicological terms this is a beneficial way to go.

During use mineral as well as biolubricants may be decomposed, and mixed with wear particles from the machine and this may lead to an increased toxicity of the waste oils. Therefore the environmental performance of used biolubricants seems in general to be poorer than of the original one. This counts especially for its ecotoxicological behaviour. However, ecotoxicity levels of artificially aged biolubricants in standard tests indicate that for used biolubricants a lower level of 100 mg/L is rarely reached, even after very long periods of use.

A large problem in denomination and recognition of biolubricants is the fact that there is no general accepted definition for *biolubricants*. In fact any lubricant based on renewable resources might be denominated as biolubricant, irrespective of the actual properties of the end product, which are in toxicological terms determined by the additives as well. At the same time however, a not renewable resources based lubricant based on mineral oil derivatives and possessing a ready biodegradability and low toxicity might use the same denomination. In practice, even products with clear undesirable environmental effects are being marketed with "green" prefaces. Therefore the need is growing to reach consensus in the lubrication world on the criteria used for biolubricants.

Several national ecostandards have been developed trying to reach an agreement at least at the national level. The actually used criteria differ to some extent, making it necessary to continue the harmonising activities. LLINCWA combined the different approaches in a classification system, by distinguishing them in a few preferential subgroups, with the feature that all described subgroups may be denominated as *biolubricant*, making group 1 the most preferred, and it is leading to the fact that the lowest group sets the limit for the *minimum criteria* (see also chapter 2). This approach gives in to national ecolabel initiatives and makes it possible to accept already (nationally) assessed lubricants as biolubricant and is, in this way, not frustrating this national initiatives. However, on the other hand it leads to the fact that the minimum requirements for biolubricants must be seen as a compromise. A critical environmental evaluation would lead to more strict criteria than those that have been agreed in these minimum requirements.

The definition of the concept of biodegradability is another problem. There are being used many different definitions to describe this property. However for classification assessments (of lubricants) only "ready" and "inherent" biodegradability are well defined by OECD standards: ready aerobic (OECD 301) and inherent aerobic (OECD 302). Due to its limited predictive value concerning the overall biodegradability the CEC test gives only limited information for the assessment of biodegradability of biolubricants.

The new EU Preparations Directive has a large impact on the chosen approach to formulate criteria for the environmental ecolabel of a number of products, including lubricants. This holds for example for the Nordic Swann, the Swedish standards and the Blue Angel. In line with this directive it is concluded that a components approach in setting criteria for biolubricants is the right way to go instead of setting criteria based on the total product. Both the ISO-standards and Dutch Vamil formulate their criteria for the total product, a pragmatic approach that avoids problems concerning confidential compositions. It is however no real stimulus for environmental product improvement, because it makes it impossible to link aquatic pollutants directly to product compositions.

## 4.9    OCCUPATIONAL HEALTH RISKS AND EXPOSURE (28-35)

Mineral base oils have a low acute toxicity by oral and dermal routes. The main effects noted in man following accidental ingestion of a large quantity of mineral oil have been irritation of the digestive tract, nausea, vomiting and diarrhoea. In the case of low viscosity products, there is a danger, if vomiting should occur, that the oil may be aspirated into the lungs, causing severe damage (chemical pneumonitis). Mineral base oils are rarely irritating to the eyes. The skin might be mildly or moderately irritated following repeated or prolonged exposure, but accidental spillage or splashing rarely causes problems. It is the solvency effect of mineral and synthetic lubricants that will dissolve the natural skin fats. Repeated exposure can cause degreasing of the skin and give rise to irritant effects like erythema (redness), oedema (swelling) and cracking of the skin. In general these more severe effects may occur when the exposure is repeated and prolonged in workers with poor personal hygiene.

One may distinguish irritant contact dermatitis and allergic contact dermatitis. Emulsifiers or soaps may influence the degreasing properties of a lubricant resulting in a higher irritation risk. But many commonly used additives like antioxidants, antiwear agents and corrosion inhibitors are classified as irritants as well. Commonly used additives with an allergenic potential are found amongst antioxidants, antiwear agents and corrosion inhibitors.

Skin problems are especially reported as a problem in using cutting fluids in the metal industry, where repeated prolonged exposure often occurs. In "LLINCWA-applications", the use of lubricants in inland and coastal water activities, the exposure to lubricants generally is of an incidental nature, during maintenance of the equipment. Although during LLINCWA in the pilot projects (see chapter 7) no systematic research on skin, or other occupational diseases has been carried out, no occupational diseases related to lubricant use were reported to the LLINCWA team.

Exposure by inhalation of airborne droplets or mists can arise due to mechanically generation of the mist (for example from pressurised sprays or contact with fast moving surfaces) or thermally generation (by condensation of vapour generated on contact with a hot surface). The inhalation of vapours or mists for short periods may cause mild irritation of the mucous membranes of the upper respiratory tract. Prolonged exposure may lead to more severe effects. A study amongst marine engineers, working in the engine room of a seagoing ship with lubricating oil (boiling point 300-700°C) and a fuel oil (boiling point 175-300°C) showed a time-weighted averaged respiratory exposure of 0,45 mg/m$^3$, 9% of the OEL (the occupational exposure limit for oil mists is 5 mg/m$^3$). Reported adverse effects were mucous membrane irritation and dyspnea (constriction) (28).

Carcinogenic properties of mineral oils are associated with their content on polycyclic aromatic hydrocarbons (PAH), but some additives are suspected in this respect as well. Mineral oils are classified as R45 (may cause cancer). Cases of scrotal cancer related to the exposure of mineral oil have been reported in many countries. The non-hygienic behaviour of carrying an oil-soaked cleaning cloth in ones trousers pocket was identified as an important route of exposure. Highly purified (or severely hydrogenated) mineral oils have a low PAH content that may result in a lower toxicity.

The documented experience with occupational exposure to vegetable oils or synthetic esters is limited, but in case of intense and prolonged skin contact similar effects as for mineral oil are to be expected. In this case the degreasing of the skin seems the most important effect.
No allergic effects are known for vegetable oils or synthetic esters. Nevertheless, additives might be used in these lubricants as well that exhibit these allergenic properties. Therefore the toxicological properties of the biolubricant are depending on the exact nature of the used additives. Vegetable oil and synthetic esters are not classified as carcinogenic.

REFERENCES

1.  H.L.Barreveld, R.P.M.Berbee, M.M.A.Ferdinandy, "'Vergeten' stoffen in Nederlands oppervlaktewater", RIZA rapport 2001.020, RIZA, Lelystad mei 2001.
2.  http://europa.eu.int/comm/enterprise/chemicals/detergents/publicconsult/reg.pdf
3.  F. Van Dievoet; "comparison of base oils and fluids from an ecological point of view : mineral-vegetal-synthetics" 7th .
4.  N.S. Battersby, The biodegradability and some microbial toxicity testing of lubricants – some recommendations, Chemosphere 41, 2000, 1011-1027.
5.  Uitvoering Strategie Omgaan met Stoffen, Voortgansrapportage, Den Haag, december 2001.
6.  C. Cecutti; "Impact environnemental de lubrifiants d'origine végétale utilisés dans l'exploitation forestière", rapport AGRICE, june 2001.
7.  Weyers A, Vollmer G, Algal growth inhibition: effect of the choice of growth rate or biomass as endpoint on the classification and labelling of new substances notified in the EU, *Chemosphere,* 41, 2000, 1007-1010
    A.  Igartua, LLINCWA report, september 2002.
8.  R. Allanou, B.G. Hansen, Y. van der Bilt, Public Availability of data on EU High Production Volume Chemicals, EUR 18996 EN/1999.
9.  EU White Paper Strategy for a future chemicals policy, EU COMM (2001) 88 final P. Benschop , RIZA report 97.024, ISBN 9036 9507 16, 1997.
10. Willing ; "Lubricants based on renewable resources - an environmentally comparative alternative to mineral oil products", *Chemosphere* 43, 89-98 (2001)
11. Willing; "Oleochemical esters - environmentally compatible raw materials for oils and lubricants from renewable resources", *Lipid*, 101(6), 192 (1999) "Les maladies professionnelles : guide d'accès aux tableaux, comité français d'éducation pour la santé" – documentation du Bureau international du Travail.
12. W.J. Bartz; "Lubricants and the environment*", New Dir. Tribol.* Plenary Invited Pap. World Tribol. Congr. 1st 1997, 103.
13. Th. van Joost and L. Reijnders, Environment and skin. The skin as a mirror of the environment (In Dutch), pp 143, ISBN 90 5352 002 3.
14. Patty's Industrial Hygiene and Toxicology, Ed. G.D. Clayton and F.E. Clayton, 4th Ed, Vol. II, Part B, J. Wiley and Sons, pp 1410-1417, 1994.
15. W. Raymond RIZA report
16. J. Guinée and all, "Life Cycle Assessment in environmentally Policy, LCA - an operational guide to ISO-standard, volume I-III, Centre of Environmental Science, Leiden University (2001).
17. F. M. Hekster, "A comparative study to environmental effects of mineral oil and vegetable oil based lubricants in open lubricating systems" *LLINCWA report.*
18. P. S. Wightman, "Environmental benefits to be derived from the use of vegetable oils in place of existing petrochemical materials", CTVO-net workshop on paints and coatings, november 1998.
19. J. Steber and C-P. Herold, " Comparative evaluation of anaerobic biodegradability of hydro carbons and fatty derivates currently used as drilling fluids", Chemosphere, 31, 3105-3118, 1995.

20. E.H.G. Evers et al., "Oil and Oil Constituents – an analysis of problems associated with oil in aquatic environment", Report RIKZ 97.032, RIZA Report 97.046, Ministerie Verkeer en Waterstaat, DirectoraatGeneraal Rijswaterstaat, RIKZ / IZA December 1997

21. ISO standard ISO/CD 15380 for lubricants, industrial oils and related products (classL) -Family H (Hydraulic systems) – Specifications for categories HETG, HEES, HEPG, and HEPR.

22. WOCB, Werkgroep Olie- en Chemicaliënbestrijding bij ongevallen op het water, "Morsingen Binnenwateren", Jaaroverzicht 2000, Directorate-Generaal Rijkswaterstaat, Ministerie van Verkeer en Waterstaat, 2002.

23. Abbink Spaink, P., *Emissies uit scheepsmotoren. Een verkennende studie van verontreinigingen uit motoren in de zeescheepvaart, beroepsbinnenvaart en recreatievaart*. STOWA-rapport 98.12. 1998

24. Van den Roovaart J.C, *Schroefasvet binnenscheepvaart, RIZA* werkdocument nr. 2001.0bbX, feb 2002

25. Van den Roovaart J.C, *Bilgewater binnenscheepvaart, RIZA* werkdocument nr. 2001.088X, jan 2002

26. Van Waveren R.H, Zeegers I,. *Doelgroepstudie en Beleidsanalyse Binnenvaart*; RIZA rapport 97.063, 1997

27. Kristin Svendsen, Bjorn Hilt, Exposure to mineral oil mist and respiratory symptoms in marine engineers, Am.J.Indust.Med. 32, 84-89 (1997)

28. Concawe, report 5/87, Health aspects of lubricants, The Hague, 1987

29. Kanerva et al., Handbook of Occupational Dermatology, Springer Verlag Berlin, Heidelberg, NewYork, 2000

30. R.K. Hewstone, Environmental health aspects of Lubricant additives, Sci.TotalEnviron, 156, 243-254, (1994)

31. P.Sartorelli et al. Dermal exposure assessment of polycyclic aromatic hydrocarborns : in vitro percutaneous penetration from lubricating oil, Int.Arch.Occup.Environ. Health, 72, 528-532, (1999)

32. Clayton & Clayton, Patty's Industrial Hygiene and Toxicology, Vol II, part B, 4th edition, John Wiley and Sons.

33. H.B.Krop, Health and Environmental Hazards of Commonly used additives in lubricants, ELGI / IVAM 2002, Amsterdam

34. Sumovera, Application of vegetable-oil based concrete mould release agents (VERA's) at construction sites and in precast concrete factories, State-of-the-art document, Febr. 1999, Chemiewinkel UvA, (IVAM)

# PART 3

# STIMULATING THE
# MARKET INTRODUCTION
# OF BIOLUBRICANTS

# Chapter 5

# Drivers and barriers for market introduction

## 5.1    INTRODUCTION

At the beginning of the LLINCWA project a market analysis has been carried out, which was subsequently refined during the course of the project. This was done for two reasons: firstly, in order to understand the low level of penetration of biolubs in the market of inland and coastal water applications as described in chapter 3. And secondly, to prepare for an effective and tailor-made marketing strategy within the project and thereafter for the different countries and market segments (which will be described in the following chapters).

Preliminary versions of the market analyses have been published in the *non-technical research report*[17] and in report *the need for further governmental interventions* [18]. This chapter provides a broad overview and summary of the findings. In the following section – section 5.2 – the overall drivers, barriers and other relevant factors for the introduction of biolubs in the market of inland and coastal water applications are presented. From the nature of these factors it becomes clear that specific market and situational characteristics are highly determining for the acceptance of or resistance to biolubs. Therefore in section 5.3 the user perspective is highlighted, by describing the way in which the overall relevant factors play a role in different market segments. The chapter concludes with a number of observations and points of departure for the development of a marketing strategy.

## 5.2    DRIVERS AND BARRIERS FOR THE INTRODUCTION OF BIOLUBS ON AND AROUND THE WATER

### 5.2.1   Introduction

In section 3.5 the relevant national backgrounds for the introduction of biolubs on and around the water in the different LLINCWA countries were described. From this description some of the drivers and barriers that play a role can be inferred. Also, from a broad range of other sources – interviews with market players, policy makers and experts, desk study and secondary data analyses – a number of actors and factors can be distilled that play a promoting, a hindering or a multidirectional role concerning the introduction of biolubs on and around the water. These actors and factors are either

---

[17]   LLINCWA non-technical research report, (1st half-year 2000), Kees Le Blansch et al. The Hague, October 2000, QA+ / LLINCWA

[18]   Le Blansch, van Broekhuizen, Novak, Ullmer, Tröbs, "The need for further governmental interventions", 30 July 2001, LLINCWA

related to market activities, to government and policy-making activities or to other influences. Table 5.1 presents the overview of these drivers and barriers. In the following paragraphs they will be discussed separately. In boxes important examples are described of the actors, factors and their resulting influences.

| | Driving forces: | Barriers: | Other (f)actors |
|---|---|---|---|
| Market barriers and incentives (see 5.2.2) | Marketing activities of specific (mainly non-oil company related) suppliers | Costs<br>Low level involvement of oil companies<br>Limited availability of suitable biolubs<br>Specific market characteristics | Performance in specific applications<br>The role of OEM's<br>Eco-labelling and related (lack of / too many) test methods |
| National and EU policies and regulations (see 5.2.3) | Legal regulation<br>Government initiatives | | Institutional characteristics |
| Other factors (see 5.2.4) | Environmental awareness | Limited awareness of alternatives<br>(Stories on) previous negative experiences | Trans-national influences<br>Nature of effective decision-maker<br>Working conditions<br>Technological developments |

Table 5.1: drivers and barriers

## 5.2.2   Market barriers and incentives

First of all, we turn to the market based barriers and incentives. We distinguish:
- Performance characteristics in specific applications (both a driver and a barrier)
- Costs
- Marketing activities of specific (mainly non-oil company related) suppliers
- Low involvement of oil companies
- Limited availability of suitable biolubs
- Specific market characteristics
- The original equipment manufacturers (OEM's)
- Eco-labelling and related (lack of / too many) test methods

*Driver and/or barrier: performance characteristics in specific applications*
Sometimes performance characteristics of biolubs can be very beneficial for specific applications. Particularly where hydraulic fluids are concerned it can be shown that intrinsically higher product prices are compensated by better performance of biolubs (less wear, improved viscosity/temperature characteristics). As a consequence a cost reduction may even be achieved. Also, for specific applications greases and gear oils can have specific advantages that allow for niche marketing (examples are given of greases for switch plates or underwater use of specific types of gear oils). It appears that for 2-stroke oils synthetic products have by far the most favourable qualities.

However, in other applications early, bad experiences with vegetable based hydraulic fluids still withhold users from applying biolubs. Stories with such experiences appear to be widespread, therewith also forming a barrier for other users to try to work with (meanwhile mostly improved, or better suited) products. Stories on other, more positive experiences are not always available.

*Barrier: costs*
Biolubs are intrinsically more expensive than traditional mineral lubricants. Moreover, smaller production quantities prevent economies of scale to occur, yet adding to the cost of their production. Reportedly they typically range from one and a half or two to four times the prices of fossil oil based ones. In particular in cases where there is no improved performance to outweigh the additional costs (for instance in the case of screw axis lubrication), these price differences may form a barrier. This barrier may also be existent in other cases, because of a lack of proper price/performance calculations.[19]

*Driver: marketing activities of specific (mainly non-oil company related) suppliers*
Another driving force for the use of biolubs consists of the marketing activities of some, mainly non oil-company related, producers/suppliers who perceive biolubs as a promising new area for competition, which allows them simultaneously to decrease their dependence from the oil industry.

*Barrier: low level involvement of oil companies*
It appears that most major oil companies (probably with the exception of TotalFinaElf and the Swedish Statoil) only have biolubs in their portfolio to meet customer demands. In the longer run it is however not to be expected that they will promote substitution of mineral products.

*Barrier: limited availability of suitable biolubs*
As a consequence of the previously mentioned factor, as well as of the vicious circle of lacking demand and lacking supply, for some applications suitable biolubs are hard to find, or at least, are hard to find with some suppliers. Often external pressure is required (like for instance in the form of governmental stimulation or regulation) to break through this impasse.

*Barrier: specific market characteristics*
Certain specific market characteristics can act as barriers. Examples are:
- The domination of small and family enterprises in inland and coastal water activities where lubricants are used. Consequently, cost awareness is high and innovation processes develop slowly.
- The long depreciation time of floating and onboard equipment, as a consequence of which replacement investments only happen after a long time. In cases where a switch to biolubs is dependent on certain equipment qualities, this characteristic severely slows down the diffusion process.

---

[19]  Often it is noted, however, that for some users costs are not the decisive factor. For these users the environmental quality represents an important quality in itself.

*Factor: The original Equipment Manufacturers (OEM's)*

In several countries and literature sources the reluctance of OEM's to include biolubs on the lists of recommended lubricants to be used on their products is mentioned as an important factor hindering the introduction of biolubs. Since the consequence of not using a recommended lubricant would be the loss of an OEM's warranty, this is seen as an effective deterrence for users to start using biolubs. Indeed it is apparent that OEM's play a pivotal role in the choice of lubricants by users. As to OEM's attitudes, however, the LLINCWA experience is less conclusive. Although the general conclusion is confirmed that the use of biolubs in inland and coastal water activities is still very low, there does not appear to be an attitude of complete dismissal among OEM's to allow for the application of biolubs. However, it is clear that a lot of – development, demonstration and testing – work still has to be done in order to make OEM's actively support the introduction of biolubs.

*Factor: Eco-labelling and related (lack of / too many) test methods*

The presence of well-known and well-accepted eco-labels for biolubs contributes to transparency in the market and facilitates users to choose environmentally friendly products. In particular from the side of producers/suppliers it is stressed, however, that the present multitude of testing methods to determine the bio-degradability and toxicity of lubricants (e.g. OECD, national, related to an eco-label) is a cost barrier for the introduction of biolubs, given the international nature of the lubricant market. On the other hand it is stressed that no (authorised) testing method is available to properly assess the environmental quality of biolubs (particularly the effects of a longer life-span of these products).

### 5.2.3   National and EU policies and regulations

Next, we discuss the drivers and relevant factors that come from government interventions. They are often seen as the more compelling drivers. We distinguish:
- Legal regulation
- Government initiatives
- Institutional characteristics

*Driver: legal regulation*

As was already shown in chapter 3, legal regulation is an important driving factor for the use of biolubs.[20] In Europe, such regulation is only in place in Switzerland and in the EU Member States Austria and Germany (several *Länder*). In the near future some effects can be expected of the implementation of the European regulation on recreational sailing in ecologically sensitive areas. Apart from this, no further

---

[20]    Another example can be found in Germany, where specific responsibilities for the processing of used or spent products are assigned to the original vendor of automotive and industrial oils.

regulation on national or European level is expected.[21] Cynical enough mineral oils have a relatively low aquatic toxicity (their environmental effects are primarily the low biodegradability and the physical staining effects) and despite the enormous yearly environmental losses and spills, mineral oils are not regarded as a priority substance in inland waters (see chapter 4). As a consequence this could be regarded as a barrier to develop a stronger environmental policy.

---

**Example I:   regulation on lakes in Germany**
Legal regulations have been installed in several German *Länder* (as well as in non-LLINCWA countries Austria and Switzerland), which forbid the use of all toxic non-biodegradable products on and around lakes (particularly Lake Konstanz (since 1982) and the Bodensee), primarily in order to protect drinking water quality.

**Example II:  EU Regulation on recreational sailing in ecologically
                    sensitive areas**
In 2001 The European Parliament adopted Amendments on *Directive 94/25/EC on the Approximation of the Laws, Regulation and Administrative Provisions of the Member States relating to Recreational Craft*, to the effect that these Member States must consider to introduce national supportive policies to encourage the use of synthetic biodegradable lubrication oils to reduce water pollution by the recreational sector. Besides, it is announced that during the review of this directive the introduction of EU wide measures shall be considered.

---

*Driver: government initiatives*
Other clear drivers are government initiatives, particularly in Germany and the Netherlands. The 'public procurement initiative' in Germany appears to have contributed significantly to the use of biolubs in publicly owned floating and fixed equipment. In the Netherlands the active involvement of the Ministry for the Environment in setting up consultation structures ('*BOMS*') and in including biolubs in the VAMIL-criteria (a fiscal measure allowing for quicker depreciation of environmentally sound investments) can also be considered a driving force for the market diffusion of biolubs.

---

[21] In some fields, however, regulation may have an influence. Think for example on the European regulation on collection and treatment of used oils (EU directives 75/329/EEC and 87/101/EEC). Also, some areas of lubricant application may require volatile organic compound (VOC) emission limits. As a point of interest, it can be noted that legislative pressure is advocated in a report on renewable materials that was produced for DG Enterprise/E.1 (Johansson, 2000).

## Some important governmental initiatives and policies

### *Environmental policy*
Environmental policies on biolubs mostly address the issue from innovative angles, i.e. not from the policy angle of chemical hazards, due to the non-priority character of this type of mineral oil pollution. Therefore policy initiatives either relate to geographic priorities (ecologically sensitive areas), product innovation policy (the Dutch VAMIL instrument) or the exemplary role of public users (Germany).

Public procurement in Germany
In Germany a call by the Federal government to all public users to apply biolubs has led to high levels of substitution in some applications – probably amounting to critical mass in this niche.

VAMIL in the Netherlands
In the Netherlands the economic VAMIL measure – granting increased investment depreciation on machines running on biolubs – has effectively altered market conditions. Dutch policy makers consider the still higher prices of biolubs as the main reason for the limited level of actual change, and additionally as a risk for falling back into the use of traditional mineral lubricants in case the VAMIL measure would no longer be in place.

### *Renewable Raw Materials policy*
Using renewable materials reach some of the principles of Sustainable Development. Agricultural raw materials that are produced with respect for the environment (no soil degradation, no water pollution,...) are renewable raw material.
Moreover, the Carbon cycle of plants is closed, so no extra $CO_2$ accumulates in the atmosphere. In this way replacing a petrol-based product by a vegetable-based one is an effective measure to combat the green house effect. The fossil Carbon remains sequestered and the recyclable Carbon of the plants is working.
To develop the use of renewable raw materials is therefore a real long-term opportunity for the industrial sector, in particularly the oleo-chemical sector producing bio-lubricants.
The conclusions on Industry adopted at the Industry and Energy Council of the European Union on 6th and 7th June 2002 calls on member states and the Commission, each within its own field of responsibility to work further on the contribution of enterprise policy to sustainable development, notably in sustainable use of natural resources and waste management to further encourage the use of Renewable Raw Materials in manufacturing industry, building on the results of the work on the "Current situation and future prospects of EU industry using renewable raw materials" developed by the Commission in co-operation with the main stakeholders concerned.

The German Market Introduction Program for Biolubricants (Germany)
The German Market Introduction Program has been established in 2000, with an annual budget of 10 million Euros. The purpose of the program is to lower market introduction thresholds for products of renewable origin. It has two main components: communication of product properties; compensation of changeover cost.
In order to receive funding, users have to choose from a list of selected products with at least 50% renewable content, not water pollutant, and good biodegradability.
In two years approx. 3.000 machines have been converted, covering a range from small agricultural tools to heavy mining equipment. More than 90% of the lubricants were hydraulic fluids. Contrary to public expectations, only five machine malfunctions or damages were reported so far.
(Source: H. Theissen, *'Experiences with the German introduction program for bio-lubricants'*; LLINCWA conference, Eibar, 2003).

The EC policy on set aside
The EC policy on set aside is mentioned as being of influence on the availability and costs of oilseed crops. As this policy is under pressure, there is serious risk for the availability and for prohibitive prices of biolubs raw materials.

*Factor: institutional characteristics*
Institutional arrangements have an important mediating effect. A clear example is the Belgian institutional three-division, the lack of a waste collection system and the probably limited implementation of an environmental protection law. These all have a clear effect on the level of biolub-use in Belgium.

### 5.2.4  Other factors

Finally, there is a number of other factors, of technological, cultural or organisational nature, that play a role with the introduction of biolubs on and around the water. We mention:
- Technological developments
- Environmental awareness
- Lack of awareness of alternatives
- Working conditions
- The trans national character of the LLINCWA issue
- The effective decision maker

*Factor: Technological developments*
Ongoing technological developments are a factor in themselves, which influence the role biolubs can play in environmental protection. In particular developments are reported that place the use of biolubs in a wholly different perspective. One of such developments is the ongoing diffusion of 4-stroke outboard engines for boats. Another development is the growing penetration of water-lubrication of screw axis on ships.

Both developments make the use of biolubs superfluous – although the precise balances of economic/environmental advantages and disadvantages still have to be made.

*Driver: environmental awareness*

From region to region and from country to country, many legal and institutional arrangements as well as activities of market actors are founded on the local awareness of environmental issues and the preparedness to take such issues into consideration. Countries like Germany and the Netherlands have typically higher scores in this respect than for instance France or Spain.

Barrier: lack of awareness of alternatives

In some cases users are not aware that there is a choice between different environmental qualities of lubrication. Lack of specialist knowledge of lubrication from the users' side, and lack of active marketing from the side of producers and suppliers, are the determining factors behind this.

*Factor: working conditions*

The quality of working conditions is reportedly negatively affected by the use of mineral oils. For example, in France it is estimated that in general 2% of industrial diseases in the workplace is mineral oil-related. However, this is mainly the issue for metalworking lubricants and cutting fluids. Though there wouldn't be a specific reason why manual handling of mineral lubricants in inland and coastal water activities would constitute less of a working conditions problem in itself, no such problem is as yet reported.

*Barrier: (stories on) previous experiences*

As was mentioned above, stories with bad experiences appear to be widespread, therewith forming a barrier for other users to try to work with (meanwhile probably improved, or better suited) products. Stories on other, more positive experiences are not always available.

*Factor: the trans national character of the LLINCWA issue*

Waterways (often) flow across borders. In the same manner boats sail through adjoining countries. In this way institutional and cultural factors are influencing one another and their mutual outcomes across borders.

*Factor: the effective decision maker*

A separate relevant issue concerns the characteristics of the person who effectively decides on the use of biolubs. Sometimes this proves to be the maintenance officer, who often has a technical orientation. Sometimes the purchasing department is the key player. Particularly in cases where these departments are separate profit centres, this is a barrier for the introduction of biolubs. On ships there is often one person in charge of most decisions – the skipper –, who also decides on the use of biolubs or other types of lubricants. In this situation, which is most usually characterised by both a lack of specialist knowledge and very limited financial manoeuvring space, little experimentation takes place and the choice for biolubs is not easily made. Besides, in some countries reference is made to strong informal networks in which these skippers operate (communicating with colleagues over board radio's). For a successful introduction of biolubs it is deemed to be of the utmost importance to take these communication channels into consideration, and to effect positive signals through them.


## 5.3    DRIVERS AND BARRIERS: THE USER PERSPECTIVE


### 5.3.1    Different considerations for different user types

In particular when we look at factors mentioned above like 'performance in specific applications', 'costs', 'specific market characteristics', 'institutional characteristics', 'environmental awareness' and 'the nature of the effective decision maker', we can safely assume that the nature and effects of these factors will be differing from market niche to market niche. Moreover, government regulations and initiatives all clearly aim at specific (sub) categories of users, thus adding to the differentiation of considerations that market players in different niches will take into account.

From our analysis of the factors and of the practices we encountered, we feel that it's important to distinguish between *at least* three types of user groups[22] (thus neglecting for the time being the variation existing within these groups):
- Skippers / inland sailing
- Recreational sailors
- Water management

Moreover, it is also important to distinguish their considerations from those of suppliers, OEM's, authorities and other relevant actors. Below, we will focus mainly on the factors that bear relevance to the three types of users, and which will come to the fore as considerations they have for using or not using biolubs. We start with a summary of the most relevant factors for these user groups in a matrix. Next we describe the impact of these factors and of the consequent considerations of these users. Finally, we briefly address considerations of suppliers, OEM's and authorities.


### 5.3.2    Overview of relevant user characteristics

We can analyse the relevant circumstances for lubricant choices of particular user groups by looking at the way in which the previously distinguished factors play a role. At first, in this section we will assess the factors themselves for the three user groups. Below, in Tableble 5.2 this is presented.

---

[22]    Other user groups, like hydroelectric power station operators, harbour authorities, dredging companies and more, are also highly relevant. Due to the smaller sizes of these groups we will not go deeper into their considerations on this place. Marketing activities aimed at them can be shaped on a case-by-case basis.

| User groups<br><br>Relevant factors | Skippers | Recreational sailors | Water management |
|---|---|---|---|
| Performance in specific applications | Biolubs offer only few advantages | Biolub 2-stroke oils are superior | Biolubs particularly superior for hydraulic fluids |
| Costs | Biolubs add (marginally) to costs | Biolubs add (marginally) to costs | Biolubs can save costs |
| Specific market characteristics | Inland sailing is fluvial / international by nature, is capital intensive with long depreciation times. Therefore dependence on OEM and slow innovation | Lubs are sold through wholesale and retail channels. Private users are diffuse group | Public bodies are more open to considerations of public interest (incl. safe operations and environmental quality) and relatively less cost sensitive |
| Institutional characteristics | Different systems of waste collection in different countries | - | Different institutional regimes for water quality protection |
| Environmental awareness | Medium | Medium | High |
| Nature of effective decision maker | Users are mostly SME's or self-employed. Therefore high cost awareness, low technical expertise, slow innovation | Users are private consumers. Impulse buying behaviour, relative cost assessment | Users are hierarchical and specialised organis-ations. Influence of power relations between functions and aspects |
| Specific regulations on lub use | None | On German lakes (DE) and in ecologically sensitive areas (EU) | None |
| Government initiatives | VAMIL in NL | None | Public procurement |

Table 5.2:   relevant factors for different user groups

In the following sections we discuss the effects of these factors for the different user groups.

### 5.3.3  Skippers

Skippers are most often sailing on boats they own themselves. Their enterprises are either family businesses or typical SME-type companies, and/or they are self-employed. The boat is the main investment, which is depreciated over a long period of time and of which the undisturbed functioning is among the prime interests of the skipper. Therefore his inclination to experiment with alternative lubs is low. Moreover, the guarantee of the OEM that the machines will also function when using the particular type of biolub is essential for his motivation.

Skippers are no experts on lubrication or machine matters. They rely heavily on past experiences, lessons learnt, and advice from parties that have proved themselves trustworthy (sometimes including suppliers). They are often not particularly well informed about the environmental consequences of the lubs they use, nor on the alternatives that exist. Communication channels are not abundant, and often neither effective, given their mobile existence (often even without addresses other than their ships) and the absence of formal organisation.

The cost sensitivity of skippers is typically high. They are repeat buyers on the market, though at the same time their business economic calculations are not as refined as to distinguish between prices and costs.

The consequence of all this is that skippers are hard to convince to try to use biolubs (including the prime need to locate them and the secondary need for assurance from OEM's and suppliers) and are critical customers when it comes to reach a positive overall assessment (based on good experiences, acceptance of higher prices and appreciation of environmental improvement).

### 5.3.4  Recreational sailors

Recreational sailors are somewhat more susceptible to environmental arguments, whereas costs are less important to them. They are impulse buyers, not repeat buyers. The recreational sailing is mostly not related to life's first necessities. It is often related to the enjoyment of nature – which adds again to the impact of the environmental argument for biolubs.

The marketing challenge for this segment lies mainly in the nature of the lub distribution channels. As they include producers, suppliers, wholesalers and retailers, a typical logic of relationship marketing and economies of scale occurs, in which environmental considerations and the dynamics of change are hard to induce. Moreover, recreational sailors lack any professional knowledge on lubrication. Therefore their knowledge of problems and of the availability of alternatives is generally low.

As a consequence, the recreational sailing market is a market with potential for biolub use, but with a complicated structure. Involvement if not incentives must be found for many wholesale and retail companies, most of which are SME's themselves.

### 5.3.5  Water management

The market of water management is the most open to biolubs. The choice of lubs in water management is mostly the result of a careful internal consideration of technical requirements and operational safety, environmental quality and costs. The latter are often well perceived, i.e. distinguished from just the price of the lub. In effect the costs of using biolubs, despite their higher purchasing price, is often recorded to be lower than the use of traditional lubs.

Environmental quality is sometimes perceived as an important criterion, given the public nature of the water management organisation and the organisational closeness to water protection boards. Moreover, sometimes government environmental policy aims particularly at the use of biolubs within its own organisations (compare the public procurement initiative in DE, or environmental protection programmes in NL).

Particularly in the larger water management organisations expertise is sometimes well developed, even through specialised functions, enabling these users to make their own judgements as to the type of lub that can and should be used. So, forerunners among public users in water management are sometimes further in their environmental demands to lubs than suppliers are able to deliver – as a consequence of which complicated discussion can arise on what is and what is not possible.

All in all, water management is the market segment with the highest potential for using biolubs – if this potential has not even already started to be fulfilled, as is the case in Germany. Barriers to overcome are mainly organisational and managerial in nature, as decisions to change lub use do – other than in inland or recreational sailing – not lie with one person.

### 5.3.6. Other actor types

Finally, there is a range of other relevant actors to be distinguished who play a role in the introduction of biolubs. Without treating these actors exhaustingly at this place, some crucial interests of these actors are pointed at to be taken into consideration when drafting a marketing strategy.

*Suppliers*
For some suppliers (particularly those who are not related to oil companies) biolubs are interesting niche products. Some of these combine biolub promotion with the promotion of high performance products, in which case the higher price of biolubs is compensated by their added value.
For others suppliers biolubs are simply products they keep in their product range in order to be able to deliver in case customers demand them. In the former case suppliers are interested in active biolub promotion and in LLINCWA. In the latter case suppliers have the tendency to give priority advice to non-biolubs.

*OEM's*
OEM's of installations and equipment only provide warranty on their products when using specific lubs after extensive tests with these lubs have been concluded with positive results. Usually they are willing to engage in such tests, however, whereas their overall attitude to biolub use is positive.

*Authorities*
The issue of water pollution through lubrication is often not priority 1. However, most authorities are easily convinced and well prepared to co-operate in initiatives to promote biolubs. Distinction must be made between authorities with responsibilities for environment, for nature, for waterways, for agriculture, et cetera. Also, with respect to several waterways supranational boards are competent (e.g. the Rhine Commission).

## 5.4     CONCLUSIONS

A number of clear overall conclusions can be drawn from the material presented above. First of all, it shows that a viable market approach must distinguish between different market segments. A differentiated marketing strategy is required, therefore. This marketing strategy must focus on

- Providing information on the environmental issue, the availability of alternatives
- Testing the applicability of biolubs in different types of equipment and installations in different sub sectors
- Proving the quality of lubrication through biolubs for a broad range of applications
- Building relationships with important actor groups and co-operate with them in effecting a marketing strategy aiming at increasing the use of biolubs.

Next, and particularly given the complicated introduction trajectory in certain market segments, the approach should also aim at improving the regulatory and institutional context. This improvement should take the following considerations into account:

- *Voluntary measures alone will not do*

Government measures that strictly aim at voluntary substitution will not provide the necessary impetus. Labels and standards are already in place and have not been effective in themselves. Also, management systems and certification schemes will mostly miss the target they aim for, given the dominance of small and medium-sized enterprises, self-employed and public lubricant users on and around inland and coastal waters.

- *Time for price incentives and regulated use*

Measures aiming directly at a price reduction of biolubs (relative to mineral oil based lubricants) are highly favourable. Tax measures can be thought of here, but also the German FNR initiative provides an interesting example of modes of financial stimulation. Indirect measures via financial stimulation of investments in machines using biolubs (like the Dutch VAMIL approach) can be fruitful in some cases. The other high potential type of government intervention is regulated use. Examples from the German lakes show the potential effect of such measures.

- *And: public procurement*

An important condition for effective government interventions is that she sets the right example herself. In the case of biolubs public bodies and authorities make a considerable part of lubricant use on and around inland and coastal waters. When public bodies seriously attempt to alter their own lubricant use, their authority increases to require other users to do the same.

# Chapter 6

# LLINCWA project design and the work done

## 6.1  INTRODUCTION

The LLINCWA project was carried out under the 5th Framework Programme within the Innovation programme of DG Enterprise. This Innovation programme, specifically focusing on the development of the *innovation concept,* turned out to be an excellent place for the further concentration of the attention on the need to introduce biolubricants in applications in and around the aquatic environment. As has been explained earlier the emphasis has been put on the non-technical aspects of the innovation process to realise a successful introduction of the biolubricants in the market.  LLINCWA made a combination of methodological *research* to collect knowledge on economic, environmental, social and organisational aspects of (acceptance of) biolubricants and combined this with pilot projects where the use of biolubricants was tested in practice followed by dissemination of the results.

Figure 6.1  The LLINCWA team on board during the Sailing Campaign in Trier

*Before LLINCWA*

The idea to organise a project like LLINCWA was born after finalising the earlier project SUMOVERA[23] within the Innovation programme of the 4th Framework programme. This project focused on the substitution of mineral oil based concrete mould release agents by vegetable release agents in the construction industry, products that are applied at (steel or wooden) moulds to prevent concrete sticking at these elements during fabrication. Mineral oil based products present an occupational health hazard and may cause soil pollution during building activities due to their toxicity and non (or low)-biodegradability, while the vegetable release agents (called VERA's in this projects) show a readily biodegradability and are non-toxic for workers. Within the commonly used very broad definition of lubricants, these release agents are also indicated as lubricants and besides some specialists for the building markets, suppliers and manufacturers of these products are the same as for other lubricants.

The SUMOVERA results turned out to be quite successful, a significant (and sustainable) increase in market acceptance could be realised for VERA's in the participating countries (an increase from 5% towards 10% in use). An evaluation in 2002, showed an redoubled use of VERA's in the Netherlands over the last two years[24].

Based on the SUMOVERA experiences, the established network and the knowledge that an enormous amount of low biodegradable, toxic lubricants are directly spilled into the aquatic environment the then Chemiewinkel of the University of Amsterdam, together with the Belgian Fina Oleochemical department developed the first ideas to set up a specific project focusing on these problems. The ideas were triggered by the use of loss lubricants in inland ships. In the open screw axe systems used in almost all older inland ships lubricants are used to lubricate the screw axe and at the same time preventing the water to enter the ship by pressing them out of the screw axe.

## 6.2    THE LLINCWA TEAM

The choice was made to focus the activities at the big Western-European water systems: the Rhine, Meuse, Schelde, Rhone etc., where aquatic pollution with mineral oil plays a major role. Coastal activities (especially harbours) were identified as an other important source of pollution with lubricants. Partners were found in the Netherlands, Germany, Belgium, France and Spain. They were selected for complementary skills and to assure representation of different market actors in the core of the project: technology and knowledge supply, users and suppliers/manufacturers. Table 6.1 shows the LLINCWA partners.

---

[23]   Application of Vegetable-Oil based Concrete Mould Release Agents, (VERA's) at Construction Sites and in Precast Concrete Factories, State-of-the Art Document SUMOVERA, February 1999, Chemiewinkel University of Amsterdam (Nowadays: IVAM Research and Consultancy on Sustainability, Amsterdam)

[24]   J. Terwoert, *Evaluatie Classificatiesysteem Ontkistingsmiddelen*, Stichting Arbouw/IVAM Amsterdam, Mei 2002.

Table 6.1 LLINCWA partners

| **The Netherlands** | Chemiewinkel UvA/ IVAM | Project co-ordinator<br>A research and advice group at the University of Amsterdam on chemistry, occupational health and environment. Merged during the project and became part of *IVAM research and Consultancy on Sustainability, department of Chemical Risks,* under the Holding of the University of Amsterdam |
| | QA+, | Questions Answers and More. A research- and policy advice group that supports organisations trying to find solutions to meet public policy demands, located in The Hague. |
| | HR | Hoogheemraadschap van Rijnland located in Leiden is a regional water management authority controlling the fresh water quantity and quality in the Western part of the Netherlands |
| **Germany** | ISSUS | University of Applied Sciences, Institute of Ship Operation, Maritime Transport and Simulation – ISSUS, a research and training facility of the Hamburg University of Applied Sciences |
| | FLT | Fuchs Lubritech GmbH, a lubricant supplier based in Weilerbach |
| **Belgium** | Valonal / Valbiom | An association related to the Faculté Universitaire des Science Agronomiques de Gembloux focused at developing non food outlets for agricultural products and rapeseed oil in particular. During the project the name changed into *Valbiom* |
| **France** | INPT-LCA | Institut National Polytechnique de Toulouse, Laboratoire de chimie agro-industrielle. LCA deals with the valorisation of both agricultural products and their by-products for non-food purpose. |
| | TFE | TotalFinaElf, a lubricant supplier based in Paris. Merged during the project from Fina Belgium to TotalFina and finally TotalFinaElf. |
| **Spain** | TRK | Tekniker Research Foundation located in Eibar is a non-profit private industry oriented technological research centre working on lubricants, tribological and environmental research. |

A key element in the project design was the establishment of *National Advice Commissions (NAC)* to give the LLINCWA activities a broad national scope. In the NAC's participate representatives of national and local governments (ministries of environ-ment), lubricant suppliers or their association, shipping organisations, technical organisations, and water management organisations.

These NAC's took care for tuning strategic decisions made within the LLINCWA project into line with the whole working field. Table 6.2 shows the composition of the different NAC's.

Table 6.2    National Advice Commissions

| | | |
|---|---|---|
| **The Netherlands** | – Ministerie VROM Directie Producten | – Ministry of Environment, DG Products |
| | – RIZA, Rijksinstituut voor Integraal Zoetwaterbeheer en Afvalwaterbehandeling | – Ministry of Transport, Directorate General for Public Works and Water Management |
| | – VSN - Vereniging voor Smeermiddel-ondernemingen in Nederland | – Association of Lubricant Suppliers |
| | – SAB Stichting Scheepsafvalstoffen Binnenvaart | – Foundation for waste disposal inland navigation |
| | – Binnenvaart Nederland | – Dutch Inland Shippers Organisation |
| | – Stichting Reinwater | – Clean Water Foundation, Environmental group |
| | – Suppliers | – Individual Lubricant Suppliers |
| **Germany** | – WSA Tönning | – User lubricants |
| | – TUHH | – Research organisation |
| | – Blohm & Voss | – User lubricants |
| | – Amt Strom- und Hafenbau | – Hamburg department of harbour and construction (authority, user) |
| | – Transport & Service GmbH & Co | – Tug company, user |
| | – Umweltbehörde HH | – Environmental authorities Hamburg |
| | – FNR, Fachagentur Nachwachsende Rohstoffe | – Agency of Renewable Resources |
| | – Bundesanstalt für Wasserbau, Küste | – Authority of the Federal Ministry for Traffic and Transport, Hydraulic Engineering and the Coast |
| | – Bundesamt für Seeschifffahrt und Hydrographie | – Federal Maritime Hydrographical Agency of Germany |
| | – Wasserschutzpolizei Hamburg | – Water pollution Control, Hamburg |
| | – Different suppliers | |
| **Belgium** | – Maritme Institute Gent | – |
| | – CEFA | – Training Centre for Skipper's |
| | – Walloon Ministry for Equipment and Transport, different departments | – |
| | – Walloon Ministry for Environment and Natural Resources | |
| | – OPVN | – Regional Office of Waterway Promotion |
| | – Port de Bruxelles | – Water management |

| | | |
|---|---|---|
| | – M.E.T. - DG Voies Hydrauliques | – Water management |
| | – M.E.T . – DG Electromécanique | – Water management |
| | – M.E.T . – DG Services Techniques | – Water management |
| | – Skippers Association | – Skippers representation |
| | – ITB | – Institute for waterways transport |
| | – Imatech | – Equipment manufacturers |
| | – Paraduyns | – Equipment manufacturers |
| | – CEFA | – School |
| | – Different suppliers | – Suppliers |
| | – Different OEMs | |
| | – EU seed Crushers' and oil processors' Federation | |
| | – IEW | – Environmental association |
| | – | – |
| **France** | – VNF, Voie Navigable de France | – French water ways |
| | – RMET-EDF | – National company for electricity |
| | – DDE | – Control of fluvial pollution |
| | – DIREN | – Regional authority for environment |
| | – IGOL | – Lubricant supplier |
| | – Hafa Lubrificants | – Lubricant supplier |
| | – Ademe | – French agency for environment |
| | – Onidol | – French oilseed inter-professional organisation |
| | – | – |
| **Spain** | – Aselube | – Association of lubricant manufacturers |
| | – Ihobe | – Agency for environment |
| | – Aenor | – Agency for normative |
| | – Asociatión de consignatarios de buques y empresas estibadores del Puerto de Bilba | – Maritime Association |
| | – Consorcio de Aguas de Bilbao | – Water Management Consortia |
| | – Asociación de industrias marítimas de Euskadi ADIMDE | – Maritime Association |

## 6.3    ACCOMPANYING MEASURES

Within the Innovation Programme *accompanying measures (AM's)* were separately organised projects, giving technical and management support to the specific innovation projects. Like the innovation projects the AM's were consortia of different organisations located in different countries with the following acronym names: Pride, Ecoinnovation, CLIP, Lifestyle, Strategi'st, Showcase[25]. All were active in the Innovation Programme to provide specialised support and LLINCWA cooperated with most of them.

---

[25]    A description of the Accompanying Measures as well as of the Innovation Projects can be found on www.showcase.com

## 6.4    LLINCWA ACTIVITIES

For the design of the LLINCWA working programme the goal oriented project planning (GOPP-method) was used. This method combines a technical and scientific input with an open brainstorm session with the potential partners in order to design a complete project programme in which all project relevant aspects are introduced and possible drawbacks and challenges are identified. A one-and-a –half day session with all partners, facilitated by the *accompanying measure* PRIDE resulted in a committed, fully informed LLINCWA-team and a thorough, detailed and workable workplan.

Three main activities were carried out by the LLINCWA project:
1. Research on technical and non-technical issues, to assure a reliable LLINCWA activity, to get a thorough understanding of the technical, the health and the environmental aspects concerning the aimed substitution of "traditional" lubricants with biolubricants.
2. Setting up of pilot projects to study the problems of introducing biolubricants into equipment (the technical as well as non-technical problems of substitution) and to find out what barriers have to be taken in practice to get the biolubricants accepted by potential users.
3. Dissemination of the gathered information to the market: to potential users and their organisations, suppliers and to relevant governmental authorities. Dissemination was carried out using newsletters, the production of informational material, the organisation of workshops and a conference, campaigns oriented at specific branch segments, discussions with branch organisations and organising pressure on governmental authorities to develop legislative initiatives to stimulate the use of biolubricants.

## 6.5    RESEARCH ON TECHNICAL AND NON-TECHNICAL ITEMS

To be clear about the research it has to be stated that *no new fundamental* research has been carried out within the LLINCWA project. Existing knowledge was gathered, interpreted and being made available for the lubricants using market. Chapters 2, 3, 4 and 5 give an overview of relevant information and data necessary for the LLINCWA team to create a thorough basis to create awareness about the existence and performance of biolubricants.

### 6.5.1   Non-technical research report

Starting point for the project was the study carried out by all the partners to get an overview of the actual *state-of-the-art* concerning the use of lubricants in inland and coastal water activities in the LLINCWA countries[26]. The study was coordinated by the partner QA+ and supported by the accompanying measure PRIDE (Gerd Paul,

---

[26]   LLINCWA non-technical research report, (1st half-year 2000), Kees Le Blansch et al. The Hague, October 2000, QA+ / LLINCWA

University of Göttingen). This report did serve as a point of reference during the whole LLINCWA project:

- It defines a first operational definition for biolubricants (ready biodegradable and non-toxic lubricants)
- It defines the major types of lubricants used in the aquatic activities: hydraulic oils, gear oils, greases and 2-stroke oils used in floating equipment, on-board equipment, fixed equipment and mobile equipment
- It gives an overview of the market position of biolubricants in inland and coastal water activities at the start of the project.
- It gives a first overview of national characteristics and circumstances that do, or might influence the use of biolubricants
- It identifies drivers, barriers and other relevant factors that do, or might influence the use of biolubricants
- It ends up with a general conclusion on the role LLINCWA has to play in their planned activities.

These general conclusions can be summarised as follows.

The first conclusion is that there *is* a role for LLINCWA. Market penetration of biolubricants for inland and coastal water activities appears to be low, both in absolute and in relative terms (as compared to other fields of application, like in forestry).

### *Potential allies*
Secondly, there is a broad array of potential allies for LLINCWA to line up with. From different countries it is reported that large numbers of actors and stakeholders show great interest in the LLINCWA project and are willing to give their support.

Some specific groups require specific attention. First of all, there is LLINCWA's prime target group, the decision maker who will have to decide to start and to keep using biolubricants. As was indicated before, it is important first and foremost to address the right decision makers (users are not always the persons in charge!), and secondly, to take into account the decision making characteristics and preferences of these persons in different specific target markets.

Secondly, there may be a different person who will actually have to use and work with these biolubricants, whose working methods may have to be compatible with the biolub requirements, whose opinion may be of considerable influence and upon whom a successful substitution may be dependent. So far no information on this has come from the country reports, for which there may be a purely research methodological reason. In the following period of LLINCWA it will be important to keep an open eye for these persons' roles.

Thirdly, the role of OEM's is mentioned. From our first exploratory round it follows that there may be a potential for success, but also a strong potential hindrance power here. Co-operation with these OEM's is therefore quite important for LLINCWA.

A final group of stakeholders that must be mentioned here are the mineral oil companies. Their market power is hard to underestimate, and their influence may go in different directions. It is important for LLINCWA to operate with caution in this respect.

### The need for information

An important task for LLINCWA to accomplish is to make more, and more reliable, information available to the different players, and, in particular, to users. Aspects of such information are:

- Drafting lists of (environmental, economic, technical, sanitary) pro's and con's of different lubricant types / systems
- Documenting practical experience with biolubricants. More specifically, (documented) success stories are needed from pilots and through demonstrations
- From a business economics viewpoint proper cost calculations and comparisons need to be provided. Such comparisons should also include all costs for functionally comparative alternatives (for instance screw axes lubricated by grease, by oil and by water, to be compared both on investment and operative costs).

In producing these success stories it is important to keep an open eye for both trans-national and national conditions that play a role in making a trial into a success. Success stories in one country may not in all cases be transferable to other LLINCWA countries due to different institutional and cultural settings. Preferably, some trials take these differences explicitly into account by comparing results across borders.

A specific point of attention in this respect is the way in which (the method along which) information is generated. A probable prerequisite for success is to deploy test methods that are accepted by most parties and that provide an adequate picture of both environmental and other product characteristics. Such a test method may not be readily and undisputedly available.

### Ranking target markets

Which markets should LLINCWA address (the issue of ranking target markets)?

- Niches where price is less important or economic or other benefits can be shown. Think of pleasure yachts, publicly owned structures and vessels, specific applications where the use of biolubricants brings along technical and cost benefits (like as hydraulic fluids).
- To introduce LLINCWA to different types of markets, it is important to aim at specific different activities like inland shipping (screw axis), harbour activities (wire ropes).
- It is worthwhile to address State owned fleets, sluices, where it is possible to point at special public responsibilities and the principle of public procurement, and where multiplier effects can be realised.

### The importance of regulation

In general, it is clear that legal regulation is, both directly and indirectly, an important driver. There are several strategies to deal with this. First, it is important (a/o. through success stories and growing market penetration) that it is proved that biolubricants *are* a feasible and environmentally superior alternative to mineral ones. The societal benefit of the introduction of biolubricants must be shown, via-à-vis the harm that is done when continuing present day lubricating habits.

Next, it will be necessary to find out in which cases voluntary/economically driven substitution is not feasible, and to present this to governments. Therefore, also target markets need to be selected which are not primarily promising, but which are important (e.g. close to ecologically sensitive areas).

Secondly, (results of) activities of one national government can be presented to others in a convincing manner by significant parties, thus showing the potential and feasibility of such measures. For that purpose, proper evaluations of – in particular – the German public procurement measure and the Dutch VAMIL-regulation are of vital importance.

## 6.5.2   Governmental Initiatives

It was identified in the non-technical study that governmental initiatives are an indispensable stimulus to create a solid base for a sustainable use of biolubricants. Therefore an overview was generated of existing governmental regulations and initiatives that stimulate the use of biolubricants in the LLINCWA-counties and the European Union or that stimulate the use of renewable raw materials for non-food purposes (like rape seed oil). Support to this study was given by the *accompanying measure* ECOINNOVATION[27].

---

ECOINNOVATION is an Accompanying Measure that intends to transfer to Innovation Projects an integrated methodology aimed at improving the environmental impact of the projects. Their main objective  is to address the promotion of a broad usage of Environmental Management Systems leading to a global improvement of the environmental behaviour of these projects; not only the projects themselves, but also the partners (enterprises and persons) comprising these projects.

One of their partners, Intechnica GmbH from Nürnberg provided LLINCWA with a study on European legislation influencing the use of biolubricants.

---

Fig. 6.2        Accompanying Measure     *ECOINNOVATION*

---

[27]  R.Beer, V. Tröbs, *Investigations on approaches for the stimulation of the use of biodegradable lubricants*, May 2001, Intechnica GmbH, Nürnberg Germany.

The resulting report *The need for further governmental* interventions strongly affirms the need for further governmental initiatives[28].

The main conclusion is that governmental regulation is the best way to stimulate the introduction of biolubricants for use on ships, waterworks and harbour installations. It points out that only few governments have taken initiatives so far. Hence a call for further-going government interventions is done.

Governmental initiatives make up for the most important market incentives so far, but they are not going far enough. Examples are highlighted in Germany, where regulation for the protection of the water quality in lakes in several *Länder* and a call by the Federal government to all public users to apply biolubricants, have led to high levels of substitution in some applications – however without providing a definitive push to the market as a whole. Similar regulation pertaining to environmentally sensitive areas in Belgium appears to be largely ineffective, due to the lack of enforcement. In the Netherlands the economic VAMIL measure – granting increased investment depreciation on machines running on biolubricants – has effectively altered market conditions. Dutch policy makers consider the still higher prices of biolubricants as the main reason for the limited level of actual change, and additionally as a risk for falling back into the use of traditional mineral lubricants in case the VAMIL measure would no longer be in place.

These practical experiences point out that governmental initiatives are very important for a successful introduction of biolubricants. In order to realise a successful and sustainable substitution, these initiatives should not restrict themselves to a limited selection of areas and should be well enforced. From all the findings it is clear that clean lubrication has a price – a price that most users will only pay if their colleagues pay the same. Technologically speaking, there is no problem to change. The real problem is the change itself, and the need to break a polluting order. It's clear that here's a task for regulators.

### 6.5.3  Lubricant emissions from water activities

It is hard to get an exact figure of the lubricant emissions from water activities, water management, inland shipping and recreational sailing related to other emissions in fresh water systems. Also figures concerning the actual (bio)lubricant use in these specific application are lacking. Therefore estimations have to be made to get an impression of the actual environmental load and the severity of the emissions from these sources.   Chapter 4 describes the results of many of LLINCWA studies to disclose data available in open literature.

To get a better understanding of the use of lubricants in inland ships a separate study on this item was performed[29]. This study revealed the important modern developments in the screw axe lubrication towards the use of *closed* screw axe lubrication, without relevant oil spill and the introduction of water-lubricated screw axe

---

[28]   Le Blansch, van Broekhuizen, Novak, Ullmer, Tröbs, "The need for further governmental interventions", 30 July 2001, LLINCWA

[29]   F.M. Hekster, *Inland shipping in LLINCWA-NL, an internship report on the stimulation of the use of bio-lubs in the inland shipping sector in the Netherlands.*  Chemiewinkel / LLINCWA April 2001 internal study.

systems. Although both systems are strongly preferable over the introduction of loss-biolubricants because all pollution is prevented, practice shows that modernisation of old systems (at the majority of the European inland ships) does not happen. Only new ships are equipped with these modern systems showing a source-oriented solution for the loss-lubrication problem of the inland ships in the long-term.

In a second LLINCWA study a thorough estimation was made of the environmental load and the resulting environmental problems caused by the use and emission of lubricants in inland ships. Beside the open screw axe lubrication systems with loss lubrication the lubrication of the rudder trunk turns out to a significant source of water pollution. A detailed description is given in chapter 4 - Environmental and workers' health aspects.

Described in the same chapter are the summarised findings from a Dutch study on the extend of lubricant releases to the Dutch waterways arising from water management, inland shipping and recreational boating and their relevance in the context of the total amount of mineral oil (and other relevant pollutants) measured in the Dutch waterways.

Identified lubrication using application in water management activities cover loss lubrication greases, hydraulic fluids, lubricating oils used for the lubrication of bearings in lock gates, turbines, generators, transformers, pumps, gears, couplings, hoist and cranes, wire ropes and chain lubrication.

Identified lubrication using application in inland activities cover: (semi) open stern tube and rudder trunk greases

Identified lubrication using application in recreational boating activities cover stern tube grease and lubricating oil for two stroke outboard engines

### 6.5.4  The biolubricant definition

One of the key elements of the LLINCWA project was to find an agreement on the definition of _biolubricants_. The industry has large interests in this market and any definition may exclude certain groups of base oils from the term _biolubricant_, which may result in a commercial drawback. At the same time however the environmental focus, especially the environmental liability is essential and a definition that covers products that may cause harm to the environment would not be acceptable as leading concept for the prefix _bio_.

An important barrier in this respect turned out to be the question: are environmental and toxicity tests carried out on the end product (the actual lubricant) or is it sufficient to carry out the tests on the components? The last option does result in an enormous saving of costs for toxicity and biodegradation tests (and of course the use of laboratory animals), because these properties of the components can be simply "translated" into properties of the end product, which as a consequence results in the fact that adaptations to products do not necessary lead to the need to perform new tests.

Nevertheless, the confidentiality about the composition of lubricants including the confidentiality on the composition of the used additive packages in these lubricants is a reason why, at least at this moment, this component-approach is not feasible. Many lubricant manufacturers do not know the exact nature of their own products due to the fact that they buy additive packages at the additive manufacturer, with well established technical data, but lacking exact chemical composition data.

Therefore, a compromise was found in the formulation of minimum requirements for biolubricants and the elaboration of a classification system (see chapter 2).

### 6.5.5   Tribological and biological analysis

Thanks to the laboratory facilities of the Spanish and the French (INPT) partners the LLINCWA team had the opportunity to perform tests on new and used biolubricants to identify their performance during their application at the pilot projects. These tests proved to be essential to generate convincing evidence on the good performance of biolubricants. The tribological tests are described in chapter 2. Toxicity and biodegradation tests are described in chapter 4.

## 6.6    PILOT PROJECTS

Pilot project were set up in all LLINCWA countries to find out technical details connected to the substitution of "traditional" lubricants with *bio*lubricants; technical barriers like compatibility of the biolubricants with the materials used in the lubricated equipment, temperature behaviour or water mix ability, as well as non-technical barriers concerning for example the loss of guarantee given by the OEM (Original Equipment Manufacturer) in case of substitution of the advised lubricant or for example overcoming a lack of motivation of a potential user (management or technicians) to change towards the use of biolubricants.

**In France the pilot projects are organised by the Laboratoire de Chimie Agro-Industrielle (LCA) of the INPT Ecole Nationale Supérieure des Ingénieurs en Arts Chimiques Et Technologiques (ENSIACET).**

**The partner**
The project partners involved in the set-up and realisation of these pilot projects to achieve
Familiarity with the use of bio-lubricants are:
- The Southwest Regional Directorate of The French Waterways (VNF).
- The Municipal Electricity Services for Toulouse(RMET).

Four different lubricant suppliers have joined these projects: Fuchs Lubritech, TotalFinaElf (TFE), HAFA Lubricants and IGOL Lubricants.

**The project locations**
The pilot projects are located over the Canal des Deux Mers and the river Garonne in the Midi-Pyrenées region.
- The Canal des Deux Mers is composed of the Canal du Midi and of the Canal Latéral à la Garonne. Due to the beauty of the scenery it traverses, it is one of the most attractive canals in Europe.
- The Canal du Midi is classified by UNESCO to be part of the 'International cultural heritage', and it is therefore the perfect place to promote environment-friendly policies through the LLINCWA project.
- The Garonne River has its source in the Pyrenees and runs to the Atlantic via the estuary of the Gironde River, close to Bordeaux.

**The pilot projects**
The Canal du Midi includes 23 locks around Toulouse. Among them, the automatic sluice of Minimes located in the heart of the city of Toulouse has been chosen for the test. The hydraulic fluid for the operation of the gates has been substituted with a bio-lubricant from HAFA.
Two more demonstration projects with vegetable based hydraulic fluids from FUCHS Lubritech and TFE are running with equipment working for the maintenance of the Canal, both are in partnership with the VNF: a dredger and an excavator.

In the other three demonstration projects, the bio-lubricants are meant for lost lubrication applications:

- the lubrication of the gear wheel of a manual sluice on the side Canal is carried out with the biodegradable from IGOL (partnership : VNF)

- the lubrication of grid cleaners of the hydro-electric power plant over the Garonne river is performed by a grease from TFE (partnership : RMET).

- The technical performances of two stroke oil from IGOL are tested on VNF engines used for the maintenance of the banks of the Canal.

**Monitoring activities**
During the two years test period 2001 and 2002, samples of hydraulic fluids are collected every 200 hours. Total acid number, viscosity, wear metal contents, magnetic particles and biodegradability are monitored. INPT-LCA and TEKNIKER have performed the analyses. The evaluation of environmental impact of bio-lubricants is carried out in eco-toxicity studies on standard species (algae, daphnia and fishes).
For greases, criteria have been identified such as adhesiveness, compatibility with seals, evolution of texture, behaviour according to temperature and waterproof characteristics.

Fig. 6.4   French pilot projects

Based on the LLINCWA market study (see paragraph 7.6) pilot projects were selected in four market area's: water management, hydro electric power plants, inland ships and the recreational sector.

In pilot projects an active cooperation of a LLINCWA partner, a lubricant user and a lubricant supplier was realised. The supplier was selected depending on existing contacts of the user with the supplier or based on the availability of specific biolubricants for the specific application.

In general many discussions (with the management, the purchaser, the technicians, supplier(s), OEM's) preceded the actual start of a pilot project. The right time for substitution has to be found (low time pressure, summer-winter), the right biolubricant type has to be selected (just fit for use, not an *over*-dimensioned or too good (and therefore too expensive) lubricant) and so on.

---

**Bio-lubricants solve technical dredging problems**
In the Netherlands about 30% of the total area and the main part of the country's economic heart are situated below sea level. Nevertheless, an increasing number of people would like to live and work in this area. One of the options to solve the related spatial problems is to reclaim new land from water. Presently, at IJ-burg near Amsterdam new land is being reclaimed by dredging and transferring sand from Markermeer, a lake north of this new area.

**Open bearings under water**
One of the dredging companies is Ballast Nedam. It is using several ships containing suction dredgers, which in the course of sucking produce large sand clouds underwater. Sand slowly precipitates on dredging installations and open bearings, which is challenging its technical performance. Ballast Nedam and other companies are looking for solutions such as closed bearings and other permanent closed seals. However, those technical solutions do not always function properly under the extreme conditions in sand clouds underwater. To keep the sand out of the open bearings a large flow of grease is necessary. Due to the huge amounts of lubricants disappearing in the water, this is an incredibly water polluting process.

**Pilot with biogrease**
A pilot project has been set up aiming to achieve a solution for reducing lubrication pollution of the aquatic environment. By using a highly biodegradable bio-grease (Green Point LCBio 1302), the flow can be increased, which can in turn automatically press the sand out of these bearings. In this project the effects of using bio-grease in underwater bearings have been studied studied.

---

Fig. 6.5  Pilot project dredging in "Markermeer"

Identified obstacles were being analysed by the LLINCWA team and answers to these obstacles were collected in a brochure "Why and how to use biolubricants" and made available for pilot projects and other interested parties[30].

Characteristic for the use and especially the testing of lubricants in the aquatic equipment is the fact that in order to get a thorough understanding of the long-term behaviour long term tests have to be performed. Information on the long-term performance, their influence on wear of the machine, decomposition of the lubricant itself (acid number), etc. are aspects that are only discovered after a long-term use; often much longer than the duration of the LLINCWA project. Of course this does not hold strictly for loss lubricants that are directly spilled into the environment during use, but information on their long-term influence on the equipment is essential as well.

---

[30]   "Why and how to use biolubricants", LLINCWA 2002
"Het hoe en waarom van biosmeermiddelen", LLINCWA 2002
"Bioschmierstoffen –warum und wie werden sie eingesetzt?", LLINCWA 2002
"Pourquoi et comment utiliser des biolubrifiant̲s̲? ", LLINCWA 2002
"¿Por qué y cómo utilizar los biolubricantes? ", LLINCWA 2002

A nearly structural problem turned out to be the organisation of pilot projects with the common commercial inland ships. Although this sector has a positive approach towards an environmentally motivated substitution, economic considerations, especially concerning working time (they work nearly day and night), were dominant in limiting the amount of succeeded pilot projects. Nevertheless contacts in this sector, especially with screw axe manufacturers turned out to be successful to stimulate the use of biolubricants.

Nevertheless in the two to three years of LLINCWA activity many projects could be monitored that at least generated strong evidence that biolubricants in general are good substitutes to their non-biodegradable and toxic (generally mineral oil based) predecessors. A selection of the pilot projects is presented in chapter 8; not all could be presented due to insufficient presentable data or due to the explicit whish of pilot projects not to be mentioned.

## 6.7    DISSEMINATION OF THE RESULTS

### 6.7.1   LLINCWA Newsletters

An important tool proved to be the distribution of newsletters, newsletters written for the users of lubricants. In total three newsletters were made, in German, Dutch, French, Spanish and English, covering the experiences and findings of LLINCWA[31].

The first newsletter describes the availability of bio lubricants state-of-the-art in lubrication, the best environmental choice, the actual use of biodegradable lubricants and hydraulic fluids by the German waterway authorities is described, successes gained by the use of biolubricants in railway systems are highlighted, the need for (lubricant) waste collection systems in Belgium is addressed, the growing of rapeseed as source for biolubricants is described and the new labelling requirements for mineral oil based lubricants (with a pictogram with the dead tree and dead fish) is explained.

The second newsletter does announce the LLINCWA Sailing Campaign, it explains extensively the use of biolubricants in equipment in the hydroelectric power plant TIWAG in Austria, it describes the initiative to develop a ranking system or classification system for biolubricants (see also 7.4.5 as well as chapter 2), an overview of the use of biolubricants at the Dutch water board Hoogheemraadschap van Rijnland is given, an article on the tribological monitoring of lubricants during use is presented, the need for further governmental interventions is stressed and it is explained how an environmental risk assessment for lubricants can be carried out.

The third newsletter reports about the success of the sailing campaign, it gives a summary of the results of the Koblenz workshop with German users, suppliers and authorities, the French workshops in Nancy and Metz, the Dutch workshops in

---
[31]   LLINCWA Newsletter nr. 1, January 2001;   LLINCWA Newsletter nr. 2, August 2001;
       LLINCWA Newsletter nr. 3, June 2002

Amsterdam, Arnhem and Utrecht, it launches the initiative for a Dutch campaign to stimulate the use of biolubricants in the recreational sector, it gives a description of the Dutch tax-reduction system Vamil for biolubricants, the use of biolubricants in Dutch dredging operations is described, French pilot projects are described, the International LLINCWA Conference in Paris is announced and the last international LLINCWA workshop in Spain is announced as well.

### 6.7.2   Marketing Plan

To structure the dissemination of the results of the LLINCWA activity and to give the publicity a larger impact a marketing plan was being developed, with the support of the Accompanying Measure CLIP.

---

CLIP supported LLINCWA in developing a marketing plan.
CLIP is an accompanying measure that assists the projects through their Marketing of Innovation methodology by reducing the marketing and technological risks that might arise during development. Through a sequence of adapted steps that go from information gathering, to a diagnosis
and the drafting of an action plan, CLIP provides help launching an unknown product/service into the market. Tools will structure the introduction of the product/service into the market and help in:
- Prioritising the market segments;
- Evaluating the impact of external actors (competitors, government, etc.) through a cartography of actors;
- Understanding and selecting these actors;
- Preparing the marketing strategy that will introduce the product into the market through the elaboration of a diffusion plan.

---

Fig. 6.6        Accompanying measure CLIP

Four market segments, relevant for the LLINCWA scope, were identified, relevant actors within these sectors were identified and a specific publicity plan was set up to address the different sectors. For the not strongly PR-skilled LLINCWA team this CLIP-support turned out to be successful.

This marketing plan contained detailed information on how to increase the awareness on the existence of biolubricants, how to address the identified actors and how to overcome possible drawbacks that might interfere with the introduction of biolubricants in the market.

The schedule for the LLINCWA action plan including identified actors and actions is presented in annex 1 of this report.

### 6.7.3  Sailing Campaign

A key element in the dissemination of "the biolubricant message" was the organisation of a "Sailing Campaign".

Fig. 6.7    The MS Reinwater ship used for the Sailing Campaign

In September and October 2001 the LLINCWA Sailing Campaign was organised. With the inland ship *Reinwater* (the educational and campaigning ship owned by the Clean Water Foundation), the information on the good performance of biolubricants was spread from Amsterdam to Strasbourg and back. The LLINCWA team sailed the Rhine, the Mosel and the Canal de la Marne au Rhine, with a bio lubricants exposition on board. At strategic locations in harbours and on board of the ship workshops were organised for inland shippers, water management groups, suppliers and governmental organisations. In Duisburg the large inland maritime fair was visited with LLINCWA campaigning material.

The workshops turned out to be a large success. Comparable difficulties to attract the attention for bio lubricants are identified in the Netherlands, Germany, Belgium and France. A most striking conclusion is the extreme low awareness about the existence, availability and good performance of bio lubricants in all different market segments. In many cases LLINCWA publications and the presentation in workshops turned out to be first time for many participants to get information about these products. After this introduction many participants took the initiative to explore the use of biolubricants in their equipment. And although the cooperation of the LLINCWA team with many biolubricants suppliers is very good, they give a good support in pilot projects, it was identified that there is a dual enthusiasm with many of them. Their activity to promote biolubricants during their normal working activities is on a very low level. Some of them (hopefully only a minority) even are used to disqualify biolubricants in favour of mineral oil based products. One of the reasons for this unwanted behaviour might be the fact that many applications still miss the approval of the OEM, the original equipment manufacturer. This emphasizes an important key action for LLINCWA to take.

In Strasbourg during a meeting with members of the *environmental committee* of the European Parliament the focus was put on governmental support for bio lubricants. Due to LLINCWA activities an amendment made by them in the European Parliament was accepted on the *Directive on laws, regulations and administrative provisions relating to the recreational sector,* that asks for the enforcement to use biolubricants in this sector. National governments are being asked to take measures that favour the use of bio lubricants in the recreational sector.

Nevertheless the need for a harmonised legal approach, also in other sectors, is an item that needs the highest attention that is brought forward in Germany as well as in France and the Netherlands. In this respect LLINCWA welcomes the highly encouraging German measure to financially support the use of bio lubricants by reimbursing the difference in price between the bio and the mineral lubricant, and on other encouraging governmental initiatives.

The Sailing Campaign was finalised with a meeting with members of the Dutch Parliaments. They as well took an initiative to stimulate the use of biolubricants in recreational sector.

### 6.7.4  Workshops

Many workshops were organised focussed on specific market actors. A common form to organise the workshops in the different LLINCWA countries was organised using a training offered by the Accompanying Measure Strategi.st in which the EASW methodology was presented.

The workshop structure used for the LLINCWA workshops was generally based on this European Awareness Scenario Workshop (EASW) idea. EASW aims at actively involving the participants in a discussion, in this case on how biolubricants can reach+ their full market potential.

*Strategi.st* integrates socio-economic-environmental aspects in R&D projects and addresses stakeholders needs and requirements with emerging needs in Innovation Projects (IPs). *Strategi.st* supports projects through the application of the European Awareness Scenario Workshop (EASW) Methodology, already validated at European level since 1994, to promote a "sustainability-based approach" of innovation. The methodology, stimulating debate, opinions exchange, new ideas generation and consensus building discussion can allow to find out joint solutions to existing problems or barriers and facilitate a concerted strategy for a "soft" technology integration. *Strategi.st* simulates by scenarios the development of the project in order to identify risks, barriers, factors of success, complementary objectives and additional requirements and help projects, through future visions, to assess how innovation can affect the external environment (natural resources, working conditions, level of employment, urban quality of life, development of rural areas) and how it can be oriented towards a sustainable implementation scheme.

Fig. 7.7      Accompanying measure Strategi.st

The involvement of the participants was generated by asking them to reflect on a background document and to one or more discussion statements prepared by the workshop facilitators. Therefore a background document was prepared with information on the different topics to be covered during the workshop. The statements provided a look into how the future might develop with regard to the use of biolubricants. The statements used were catchy and short sentenced. Examples of statements used in different LLINCWA workshops: "*In 2010, biolubricants are used for all applications in inland shipping*" or "*An EU wide eco-labelling is a powerful and necessary market incentive for biolubricants to gain their full market potential*" and "*Within the next five years a EU wide ecolabel will be launched*".

**Workshop Koblenz**
As part of the Sailing Campaign, in October 2001, a workshop in Koblenz was organised to provide a forum to discuss the state of the art in replacement of mineral oil by biolubricants in inland shipping, hydroelectric power stations and hydraulic engineering.

The addressed actors were biolubricant suppliers, users and operators of hydraulic installations and authorities. Some highlights are summarised below for each topic elaborated in the presentations and discussions focussing on successful biolubricant applications.

*Legal aspects, research and public initiatives*
In Germany legal liability is based on the assumption that –in case of environmental contamination– plants, which are capable of emitting pollution are the polluters. This means the reversal of the burden of proof for operators. In case of an oil-accident operators should cooperate with the responsible public authorities to ensure adequate measures and increase transparency between all parties involved.

*Added value of biolubricants in hydraulic engineering and hydroelectric power*

The Tyrolese Hydroelectric Power (TIWAG) attaches great importance to environmentally sound operation and construction. Replacement of mineral oil by biodegradable non-toxic lubricants is going on in several hydroelectric power stations. Starting in the planning phase, thorough engineering and exact dimensioning of components ensures economic feasibility. Cooperation in partnership with suppliers during the whole lifecycle of the plants ensures high ecological, economical and technical quality standards.

*Applications in inland shipping: risks, costs and benefits*

The LLINCWA pilot project 'Hubinsel Annegret' is a large working pontoon operating with biodegradable hydraulic fluids and biodegradable greases in loss lubrication. The additional cost for biodegradable lubricants are an important factor of systems security and ensure minimised environmental risks. High security standards fulfil legal requirements and contribute to trouble-free and economic operation of the pontoon.

*Conclusions and recommendations*

The most important identified obstacle is a lack of information about the benefits of biolubricants. The most promising actions for increasing the market share of biolubricants are:
- To increase information exchange (through an information pool / News Group)
- To bundle activities on public relations, technical and scientific knowledge
- To install clear and harmonised legal requirement all over Europe.

**Workshops Nancy & Strasbourg**

As part of the Sailing Campaign, in Autumn 2001, INPT organised two workshops, which were attended by important stakeholders active in the supply chain of biolubricants – i.e., manufacturers (Fuchs, TOTALFINAELF, HAFA), retailers (Beyel, IGOL), a user (Port Autonome de Strasbourg, VNF), the BfB oil research institute, the French Oilseed Interprofessional Organisation ONIDOL and the French Agency for Environment and Energy Management (ADEME).

During the workshops several topics were covered. INPT presented the French pilot tests. Fuchs and BfB described the technical and environmental performance of their products in practice. ONIDOL addressed the topic of 'standardisation and eco-labels'. ADEME highlighted the French policy on the promotion of biolubricants. The promotion of bio-lubricants is part of the AGRICE programme (Agriculture for Chemistry and Environment) that focuses on new uses and enhanced value for agricultural products. AGRICE is set up by the French government and is managed by ADEME. The discussion concentrated on the environmental and technical aspects of lubrication and the role of additives. Additives, necessary to achieve the required technical performance level, often have a negative impact on the environmental performance of lubricants based on fast biodegradable and non-toxic base fluids like vegetable oils and oleo chemical esters. While significant results have already been achieved in the field of fast biodegradable and non-toxic additives, this work must be maintained and amplified in order to identify the full range of environmental sound additives. The participants stated that the general public is not fully aware of the environmental burden lubrication can cause. Another issue put forward was the need for a broad consensus on testing methods for assessing the environmental effects of lubricants. Once these standards are established a broadly accepted eco-labelling scheme could be established. Subsequently, the lack of legislation and regulations in favour of bio-lubricants was discussed. Participants questioned whether legislation is the best approach to promote bio-lubricants, because of anticipated opposition by industry wanting to preserve a maximum degree of freedom and flexibility. They stated that effective regulation should fairly direct all actors involved in the life cycle of lubricants to take responsibility for the impacts to human health and the environment resulting from production, use and disposal of lubricants. Finally, the issue of the higher costs of biolubricants was dealt with as well as how these can be balanced, taking into account the technical advantages of biolubricants and their superior environmental performance.

### Workshops in Amsterdam, Arnhem and Utrecht

As part of the Sailing Campaign, in Autumn 2001, three workshops on the use of biolubricants were held in the Netherlands. They were highly attended and well evaluated. Some highlights are summarised below.

*Greater incentives needed to promote biolubricants*

Suppliers, governmental authorities, policy makers and Original Equipment Manufacturers (OEMs) attended the workshop. The absence of skippers at this workshop demonstrated that biolubricants are not a high priority for them and that LLINCWA should put extra effort in grasping their attention and interest. In one of the four presentations representatives of the Dutch foundation for the collection of waste materials from inland shipping (SAB), joined with QA+ Consultancy and Advise, to elaborated on the need for government interventions to promote biolubricants. It was stated that biolubricants for inland shipping purposes are currently too expensive and may only succeed in entering the market with the support of a regulatory framework like the VAMIL tax incentive. In the ensuing discussion several attendees commented on the need for additional governmental initiatives. Regulations should be simplified and harmonised across Europe to make them effective. After all, VAMIL applies only to closed installations in the Netherlands. The use of biolubricants in open systems is currently not being encouraged by governmental measures. Another issue addressed was that governmental authorities apparently are not fully convinced of the environmental benefits of biolubricants. LLINCWA has an important role to play in:

- Generating and communicating reliable information on the environmental benefits of biolubricants
- Providing a platform to discuss the need of regulatory action at national and Community level.

*Introduction of biolubricants feasible without technical problems and extra costs*

Suppliers, OEMs, policy makers and (mainly) technical engineers from water boards attended the workshop. Two water board maintenance engineers (of Hollands Kroon and Hoogheemraadschap van Rijnland (HHR)) demonstrated that the 'change over' from mineral oil based lubricants to biolubricants is most of the times cost neutral. It may even provide some cost savings if one factors in higher tool life as well as energy and noise reduction due to the special properties of biolubricants. The industry representatives presented an impressive long record of successful applications of biolubricants in installations operating close to waterways. In the discussion it was stressed that the OEM should always be consulted before using biolubricants in installations. The most striking conclusion was that biolubricants get little attention at the water boards. Many participants pleaded for incentives to promote the use of biolubricants. Others pointed at the role that the Association of Water Boards can play in terms of facilitating individual water boards in their efforts to switch over to biolubricants.

*Drafting plans to include recreational boating into LLINCWA's area of action*

At the official closing of the sailing campaign, representatives from NGO's, industry and members of the Dutch Parliament were brought together. They reviewed the experiences with biolubricants and their role in reducing diffuse water pollution and identified policies that can be adopted by authorities in order to promote their use, especially by recreational boating. In the discussion a representative of Waterpakt, an umbrella organisation of environmental NGO's active in the field of water protection, described recreational boating as an underestimated source of diffuse water pollution. Also the technical advantages of biolubricants (by a representative of TotalFinaElf) and the technical applicability of biolubricants from a users point of view (by HHR) were addressed. In the end, members of parliament acknowledged the importance of promoting biolubricants and committed themselves to provide political support for the mission LLINCWA.

**Belgian Sailing Workshop**

Under the heading *"Inland Water and the Environment"* a Belgian LLINCWA workshop was organised on board of *Equisonnances* ship, on the *Ath to Blaton* canal. An excursion on the canal gave the opportunity to see the locks that are greased with biolubricants (LLINCWA demonstration). About 60 people attended the conference and the excursion It was organised by LLINCWA in cooperation with the group *Inter Environnement Wallonie* (an environmental oriented association) and the *Office de Promotion des Voies Navigables* (the office for the promotion of waterways), a public institution including skippers, public managers of waterways, pleasure boats/harbours associations and management. The workshop was targeted at a large public concerned by transport, lubricants, water and environment and the formulated aims were

- To promote the waterway as an environmental friendly transportation mean
- To show the need to keep (to get) the waterways clean
- To promote the biolubricants in water activities

**Workshop Mannheim**

The concept/program of the sailing campaign, beneath workshops on board, also included the organisation of info-seminars, of which one was held on the premises of FUCHS PETROLUB AG, Mannheim, when the ms „REINWATER" anchored there on 5th October 2002. About 80 participants, for example customers and interested persons from industries, water management and related authorities took part in the seminar, furthermore also the local press was attending to issue an article about the event.

The program consisted of two sessions, one part in-house which mainly dealt with presentations and papers about bio products, application examples and know-how exchange between the speakers and the guests, the second part, held on board of the ms „REINWATER", included the visit of the campaign-ship combined with an exhibition and the diffusion of information material about the LLINCWA project, its targets, activities etc.

**Workshop in Toulouse**

INPT and Voies Navigables de France (French Waterways) organised a workshop in Toulouse the 11th September 2002 which were attended by important persons from varied activities in connection with biolubricants field: manufacturers (Fuchs Lubritech, TotalFinaElf, Condat, Cognis), users (VNF, EDF Electricity of France, RMET Régie Municipale d'Electricité de Toulouse, SHEM Société HydroElectrique du Midi), the French Oilseed Interprofessional Organisation ONIDOL and the French Agency for Environment and Energy Management (ADEME).

The workshops allowed presenting results of the French pilot tests in progress on the Canal du Midi and on the river Garonne. Among the 23 locks around Toulouse, one of them, the automatic sluice of Minimes located in the heart of the city of Toulouse has been chosen for testing hydraulic fluid for the operation of the gates. Two more demonstration projects with vegetable based hydraulic fluids are running with equipment working for the maintenance of the Canal, both are: a dredger and an excavator.

In others demonstration projects, the bio-lubricants are meant for lost lubrication applications.

Technical and environmental analysis performed by INPT and Tekniker demonstrated excellent behaviour of biolubricants during pilot tests and convinced all participants about qualities of these products.

The discussion that followed was about technical and environmental performances of biolubricants but also their limits, the environmental benefit for the energy management, the process of biolubricants and the questions about OEMs guaranties in case of use of biolubricants.

Suppliers presented their current research work, their futures products and the potential development of biolubricants.

EDF and SHEM showed their intention to test these environmental friendly products in their equipments and the experiences carried out in Toulouse were of particular interest for others VNF regional divisions.

This technical meeting gave also the opportunity to VNF to develop its environmental policy and all participants were in expectation for the Paris Conference and the final report to measure the outcome of the LLINCWA project.

**Workshop Eibar**

In the very last month of the LLINCWA project, an international workshop on bio-lubricants 'From development to the market' was organized in Spain. LLINCWA and the Virtual Tribology Institute (VTI) the Consortium jointly organized this workshop.

In the context of the VTI, the performance of the bio-lubricants were analysed from the perspective of friction and wear; also new topics were discussed. In the context of

LLINCWA, the knowledge acquired on biolubricants in inland and coastal activities were disseminated. The joint action of VTI and LLINCWA did allow the development of a general approach to the possibilities of using biolubricants in the market and it gave the opportunity to discuss the needed future research and development. This workshop did mark the end of the LLINCWA project, discussed the conclusions and launched the initiative to organise a follow up of the biolubricant R&D activities within the frame of the 6th European Framework Research Programme. A follow up will be fitted into the new philosophy on integrated projects.

As conclusions of the workshop, it was analysed that for the successful introduction of biolubricants in at least a 30% of the market the following needs for research and development are required:

*Raw Materials*
- To assure enough quality of raw materials (viscosity, constant quality, capability, scale up, reduce cost)
- *Additives*
- Eco-additives, designed to work with biodegradable synthetic oils, structural-function relationship, guide for facilitate the choice of additives, formulation with lower additive content.

*Standards*
- To assure availability of standards to evaluate the Biolubricants (biodegradability, toxicity, friction, wear) in combination with materials and coatings.

*Labelling*
- To establish a common European criteria to classify lubricants for different applications (cutting oils, greases, hydraulic, gear, engine oils)
- To establish a database of existing products.*Development and Demonstration*
- To evaluate friction, wear, noise and vibration reduction, compatibility with materials and coatings, oxidation performance and temperature resistance.
- To demonstrate those Biolubricants that are already available

*Condition Monitoring*
- To increase the lifetime of Biolubricants
- To reduce oil change interval
- *Recycling*
- To assure a correct recycling and treatment procedures to optimise the final end of Biolubricants.

*LCA and environment*
- To evaluate the impact of RRM in the decrease of $CO_2$, and establish the complete LCA analysis.
- To know the effect of the diffusion leakage in the ground water.*Creating consumer demands, warranties and guaranties*OEMs should assure the suitability of the biodegradable fluid, and be prepared for factory fill if specified.*Cost and competitiveness*
- Scale up of the production of Biolubricants
- Tax incentives should be considered from Government*Market introduction*
- To coordinate national and European activities for market introduction programmes
- *Training and dissemination*

- To raise public awareness about the environmental benefits of the use of biolubricants and RRM

*Legislation*
- The adequate legislation frame and moment needs to be defined


### 6.7.5  Recreational boating campaign NL

Two-stroke engines and open stern tubes of pleasure motorboats form an important source of water pollution in Europe. An estimated 30% of all fuel and oil used in two-stroke outboard engines ends up in water. Stern tubes of relatively old leisure motorboats form another source of environ-mental concern, because these systems are usually open and grease lubricated.

There are approximately three and a half million pleasure boats in EU countries, including Switzerland. The majority of them are motorboats. Taken individually, their impact may be small; collectively it is a major concern. LLINCWA is responding to this concern with a special campaign targeted at the recreational market. For the moment the campaign was focused on the Dutch recreational market.

*Application of lubricants in leisure boats*

Leisure motorboats may pollute both air and water. Among other types of emissions, leisure boats discharge lubricants into water. Lubricant emissions from leisure motorboats come down in two categories: emissions of lubricating oil from two-stroke outboard engines and emissions of grease from lubricated stern tubes. Most outboard engines are two-stroke petrol units using an oil/petrol mix, although four strokes are slowly growing in popularity. Two stroke engines use a method of combustion, which results in unburned lubricating oil and fuel discharges into water. Stern tubes encountered in pleasure motorboats may be grease or water lubricated. A rule of thump when estimating the amount of vessels with grease lubricated stern tube systems is the following: motor craft fabricated before 1980 is most likely to be fitted with an open, grease lubricated stern tube system. Given the long lifetime of vessels one may conclude that a considerable amount of leisure craft still in use is fitted with grease lubricated open stern tubes.

*Recreational boating in sensitive areas*

The lubricant emissions of recreational motorboats are estimated to be small compared with the emissions of other, mainly onshore, sources. However, this global estimation can be misleading. Recreational boats are generally used in pleasant weather conditions and mainly at weekends and in environmentally sensitive areas. Especially on bright summer weekends, the amount of emissions in a boating area cannot be neglected. Furthermore, summer is the most sensitive reproduction period for aquatic organisms.

*Biolubricants for recreational crafts*

Biolubricants can alleviate the environmental problems caused by recreational motorboats. While biolubricants are ready available through manufacturers and wholesale distributors they are hardly marketed and difficult to find on the shelves of retail shops in marina's and bunkering stations. At least, this is the situation in the Netherlands. The reasons for this are multiple, but one thing is evident: the supply of all retail shops drives costs up, especially because the demand for the product is low due to the fact that motorboat owners are simply not aware of the existence of biolubricants.

**Biolubricants for the recreational sector**

The Dutch LLINCWA team has carried out a communication campaign aiming to promote the use of biolubricants in the Dutch recreational market. For this campaign LLINCWA has jointed forces with Waterpakt, an umbrella organisation of environmental NGO's active in the field of water protection, and with both wholesale dealers like TotalFinaElf, Fuchs and Castrol and a retail dealer (Greenway Products).

The most important actors in this sector have been identified. A good number of them committed themselves to cooperate with our efforts. The actors that cooperated with LLINCWA within this campaign were: Dutch Driving, Biking and Walking Tourist Bond (ANWB), the traditional professional sailing business BBZ, the Association of Dutch Water Boards, the Ministry of Transport, Public Works and Water Management, the Dutch Fishing Association (NVVS), the Dutch Water Sport association (KNWV), the Dutch Water Ski Association and the Dutch Yachting association (HISWA). They all showed their readiness to collaborate by providing their logo for a LLINCWA / Waterpakt poster calling the consumers in the water recreational sector to use biolubricants. The poster was distributed by LLINCWA and all the cooperating organisations to their members. Besides a sector-specific brochure has been made. The brochure was made available at the HISWA Amsterdam boat show 2002 at the stand of the ANWB. By this activity LLINCWA became well known in this market segments. There is the potential to continue our efforts for the promotion of biolubricants in the water recreational sector also after LLINCWA has finished. Plans have been drawn to set up project with Dutch local authorities and environmental NGO's.

### 6.7.7    Involving other countries

Strong efforts have been put to involve new countries into the LLINCWA activities, especially the UK, Greece, Italy and the accession countries Poland and Hungary. The initiative was taken by British Waterways, which was informed about the beneficial properties of biolubricants but could not identify any activity so far in the UK where these products were used in water management equipment. Support was asked from asked and after discussions with the European Commission an initial proposal was developed for an interim extension of the LLINCWA-team to the UK and Greece. Unfortunately interim extensions were not allowed in second instance, leading to the fact that the mentioned countries could not be involved in LLINCWA as formal partners. So a large proposal was made for the 3rd Call for Proposals of the Innovation Programme in spring 2002 to extend LLINCWA to the mentioned countries and to involve more user groups, especially water boards and shipping organisations. Many of the LLINCWA findings and experiences were assimilated in this proposal. This proposal however, with the acronym Aecoline (Aquatic environment compatible lubricants in the environment), was rejected.

Nevertheless, dissemination of the LLINCWA findings to third countries was organised by presenting results and experiences at different international fora and conferences.

# PART 4

# LLINCWA'S PRACTICAL
# EXPERIENCES

# Chapter 7

# Practical experiences with bio lubricants

## 7.1. INTRODUCTION

This chapter presents the practical experience with biolubricants built up during LLINCWA. More specifically, it discusses examples of "biolubrication" as it applies to equipment encountered at water management facilities like water treatment plants and pumping plants, hydroelectric power plants as well as the lubrication of shipboard machinery. These are typically systems susceptible to the possible release of oil into waterways due to operational characteristics (loss lubrication) or through water wash off, incidental leakage due to damaged seals, accidental discharges from failed equipment or operational errors.

In order to gain practical experience field projects (pilots) have been set up by LLINCWA in different sectors and in different LLINCWA participating countries. A pilot project is often the last major milestone prior to substitution. It is a part of the effort to convince organizations of the value of biolubricants. The results of pilot projects can be shown to decision-makers as evidence of the biolubricants' value and can provide a tangible way of communicating the potential of biolubricants to sceptics within the organization. Besides, pilots are useful for verifying estimates of costs and benefits. In summary, pilots provide a range of reduction of risks associated before final commitment to final use is made

The main support and assistance provided by LLINCWA under these projects has been threefold:
- informing users (operators and decision makers) on the possibilities of biolubricants and providing motivation to actively participate in LLINCWA activities;
- monitoring the performance of biolubricants
- bringing together and establishing cooperation between users, suppliers of biolubricants and sometimes OEMs.

The participating countries were Germany, the Netherlands, Belgium, France and Spain. The sectors primarily involved were water management, hydroelectric power, inland shipping and professional recreational shipping. Additionally, information was gathered on the use of biolubricants in various other sectors and on an number of existing field tests.

**Selection of products**

The tested biolubricants involved mainly synthetic esters, often based on renewable raw material. In other cases vegetable oil based biolubricants and sometimes polyglycols were used. A great number of suppliers was involved (Aral, Castrol, Shell, Kleenoil Panolin, Fuchs Lubritech, TotalFinaElf, Mobil and many others).

Most of the pilots were set up before the LLINCWA classification system (see chapter 4) was finalized. Consequently, the identification and selection of preferable products to be tested in each pilot was not done on the basis of the environmental and health criteria involved in the classification system. Users were allowed to select the best available product for their needs and balance the available environmental information and information on health aspects along with traditional factors such as price, performance and availability of products.

A follow-up with manufacturers of some of the products tested revealed that some products tested in pilots (and described here) do not qualify as biolubricants according to the minimum criteria set forth in LLINCWA classification system. Still the tested products are more preferable than the conventional ones from environmental point of view.

## Applications

Much of the equipment involved in the LLINCWA pilots was grease lubricated.  In most of this equipment the greased components were in contact with water. Examples include manual locks, racks in floodgates of hydroelectric power stations, stern tube lubrication and bearing lubrication. Many pilot applications deal also with hydraulic systems as e.g. cranes and decks hydraulics, hydraulic drives in sluices. Often the systems work in a humid environment and within a wide temperature range. Other applications are turbine oils in hydroelectric power and gear oil in gear boxes.

## Technical and non-technical aspects

Factors affecting successful substitution of mineral lubricants were identified and analyzed during the pilots.  The technical and non-technical aspects of substitution and a set of steps for a successful replacement of mineral oil based lubricants are thoroughly discussed elsewhere in this report. What follows is a short summary of the main findings acquired from the pilots.

Although the substitution process differs from one organization, market segment or national market to the other, the main actors involved are in all cases very similar: users, suppliers and equipment manufacturers.

Users may be categorized on the basis of their size in small and large enterprises. Besides, users may be classified as private and public-owned enterprises.

Conversion to biolubricants may require special considerations, measures, or adaptations to the system. Accelerated fluid degradation at high temperature, change of performance characteristics at low temperature, possible new filtration, maximum allowable water content in the lubricant and maintenance requirements should be investigated carefully.

User's prior experiences with bio lubricants may therefore be an advantage. On the other hand bad prior experience with products that proved to be unsuitable are an obstacle. No prior experience at all may as well result in resistance to change.

Machinery manufacturer's approval is normally required when switching to biolubricants. Lack of manufacturer's approval has often been mentioned as a major barrier for substitution of mineral oil. However, the situation has much improved the last years. Nowadays most Original Equipment Manufacturers approve at least one biolubricant for their systems.

The choice between vegetable oil based, synthetic materials or polyglycols is often governed by the prevailing power and the operating temperatures, with synthetic esters and polyglycols being preferred for working in sub-zero temperatures or temperatures of more than 90°C, and vegetable oil-based fluids in between these limits. Vegetable oils are additionally not being used when water contamination of the system is an issue. Vegetable oils not only tend to absorb water, but are also prone to hydrolysis, or decomposing in a chemical reaction with water when heated, which can result in decomposition and degradation. However, at more moderate temperatures these fluids can operate at high pressures with low wear and high viscosity for extremely long periods, provided the system is kept free from water contamination.

In large companies lubricated equipment is designed and maintained by specialized technical personal that has the competence to support the substitution process and make educated application of biolubricants while small companies have to get along without this specialized technical knowledge. Small enterprises have to rely heavily on their practical experience and the expertise and personal involvement of the lubricant supplier who may not always be as enthusiastic when it comes to promoting biolubricants. One plausible interpretation for this reserved attitude is that biolubricants are relatively new products when compared to the mineral oil based ones. As relatively new products, biolubricants require a more structured approach and more intensive consultative services when introduced on company level than traditional mineral oil based products. Since most suppliers handle in both bio- as mineral oil based lubricants, to choose for actively promoting biolubricants is a risk that not many are willing to take.

Public owned enterprises are generally more sensitized to the need to introduce biolubricants.
Besides some public procurement directives do include environmental purchase criteria for lubricants leading to a relative big market share for biolubricants in public owned sectors, as it is the case in the German water management sector.

In principle well performing biolubricants are available for a wide range of applications. However, there are national as well as market segment specific differences. In general, one may state that the lubricant market is a demand-driven market. Lack of awareness and simply lack of knowledge on the existence of biolubricants (from the side of potential users) results in low market shares. In the water recreation sector for example, biolubricants are hardly to find on the shelves of marina shops. Consequently potential users are not asking for biolubricants unless they become informed on their existence and suppliers are not doing any special effort to actively promote biolubricants due to the low market demands and the fact that they instead provide their customers with mineral oil based lubricants. LLINCWA has tried to address this typical chicken and egg problem by informing potential users on the possibilities of biolubricants.

At present there are mainly big companies that dominate the lubricant suppliers market in Europe and there are but a few independent small companies and cooperatives that are commercially producing biolubricants.

## 7.2.   OVERVIEW OF THE LLINCWA PILOTS

Environmentally friendly lubricants have been tested / surveyed in 35 different equipment. The pilots are presented below, grouped by sector segment and country.

### 7.2.1   Water management / The Netherlands

#### *Hoogheemraadschap van Rijnland*

Rijnland is one of the biggest water boards in the Netherlands. Water boards are local governmental bodies, which have the single task of managing the quality and quantity of water in their individual areas. Since 1200, Rijnland has controlled about one quarter of the surface waters in the provinces North- and South Holland in the western Netherlands. This is a region of over 1000 square kilometres between Amsterdam, Ijmuiden, The Hague and Gouda, called "the heart of Holland". About 160 polders, lakes, rivers and canals are located in this area.

Rijnland is one of the LLINCWA partners with 10 year of experience in the application of biolubricants. Their involvement in the LLINCWA project has enabled Rijnland to extend its work on this field with new applications, new suppliers and new products. Since Rijnland started to apply biolubricants in 1992, the total amount of used lubricants decreased by 15%, while the total lubricated equipment increased more than 25%. In 2000 the share of biolubricants in the total purchase of lubricants was as follows:

| | |
|---|---|
| Greases | 52% |
| Gear oils | 47% |
| Hydraulic fluids | 32% |

The highest volume use of products at Rijnland are synthetic esters from vegetable oil (rape seed) and vegetable oil based lubricants manufactured by ExxonMobil Lubes Industry, Castrol, Shell, Fuchs and Wits used primarily for bearing lubrication. Other applications include gear oils and hydraulic fluids.

The higher costs associated with the purchase of biolubricants can, according to Rijnland, be lessened or even offset by the longer lifetime of biolubricants. Through cost-effective lubricant policy Rijnland currently saves about 10.000 Euro a year. This is achieved by applying longer lubrication intervals (if technical feasible) based on monitoring of the lubricant during use instead of changing oil after a preventative fixed deadline.

Besides, temperature reduction, energy reduction and noise reduction associated with the use of biolubricants as well as low emissions and significantly reduced burden on the environment more than compensate for the higher purchase costs.

*Grease*

About the use of biolubs the users report problem free operation since 1994. The first experience with biolubricants was with Mobil UF.2 (based on pure vegetable base oil) which replaced the product Mobilplex 47 (based on mineral base oil). For equipment with high rotating speed, operating temperature and power the choice is then EAL.102 based on synthetic ester.

After years of successful biolubricants use at many locations with different applications a problem occured. Due to the merging of Mobil with Exxon it was not possible to maintain  the exact composition of UF2, resulting in problems with the pump ability.  This problem occurred especially in grease pumps with long pipelines and mobile cylinders. Resin formation and adhesion phenomena around the piston with cylinder were the result. The problem was solved by substitution towards an alternative biolubricant, a fall back towards conventional lubricants was avoided.

*Hydraulic oil*

Users report problem free operation since 1994. The first experience with biohydraulic fluids for the hydraulic systems was Mobil EAL.224H. Later on after when the pump cylinders functioned too slowly in wintertime in some applications a better choice was found to introduce a hydraulic fluid based on a synthetic ester.

*Lubricating oil*

For gearboxes users report problem free operation since 1994. The first experiments were performed with an environment friendly product (long time used) and not a LLINCWA defined biolubricant. Additional advantages of these products turned out to be a reduction in use, lower operating temperature, and decrease the use of energy. After 1999 in the LLINCWA period the choice was mad for a real biolubricant with for example a better FZG value (load stage) for micropitting and a reduction of noise.

Some field tests in Rijnland, started during LLINCWA, are presented here bellow:

**1**
**Location**:       Boskoop water treatment plant
**Description**:    underwater bearing / screw jack
**OEM**:            Spaans Babcock
**Biolubricant**:   Mobil EAL 102

**Evaluation**:
The user reports problem free operation since 1999. Mobil EAL 102 has replaced in this application Mobiplex 47, a mineral oil based grease. Mobil EAL 102 is a lithium complex grease based on a synthetic ester derived from rapeseed oil. It is formulated to deal with operating temperatures ranging from –20 to 100°C, is resistant to water wash-out and has good EP characteristics. It meets the environmental and health requirements of the Clean Lubrication Project in Gothenburg. Besides, it is a BG1 grease according to the LLINCWA classification system. BG1 is the category of the most preferable greases.

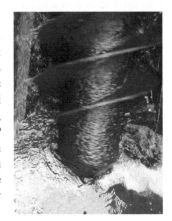

**2**
**Location:**       Waarderpolder, Velsen, Zandvoort
**Description:**    underwater bearing /open system / screw jack
**OEM:**            Spaans Babcock
**Biolubricant:**   Castrol Biotec HM

**Evaluation:**
User reports problem free operation since 2001. After 3/4 of a year the machinery was completely disassembled and inspected. Particular attention is focused on unusual damage and wear. No unacceptable signs of wear were observed. Castrol Biotec HM is a lithium complex grease based on a synthetic ester derived from rapeseed oil. It is a NLGI – 2 grease formulated to deal with operating temperatures ranging from –25 to 100°C.

**3**
**Location**:       Halfweg pumping station
**Description**:    3 gear boxes
**OEM**:            Flender
**Biolubricant**:   Castrol Tribol 1418 / 460 (gear oil)

**Evaluation**:
User reports problem free operation since 1999. Oil condition was evaluated after ca. 8000h. The samples analysed show aging complying with the normal tolerance for used oil. The oil was still usable for further use. Castrol Tribol 1418 / 460 is gear oil based on a synthetic ester derived from a vegetable oil. It is formulated to deal with operating

temperatures ranging from –25 to 90°C. The oil has good EP properties. Castrol Tribol 1418/460 is more than 80% biodegradable according to CEC L-33-A-93.

**4**

| | |
|---|---|
| **Location**: | Halfweg pumping station |
| **Description**: | Underwater bearings |
| **OEM**: | Spaans Babcock |
| **Biolubricant**: | Mobil EAL 102 |

**Evaluation**:
User reports problem free operation since 1999.

**5**

| | |
|---|---|
| **Location**: | Halfweg pumping station |
| **Description**: | Grid cleaner |
| **OEM**: | Landustrie |
| **Biolubricant**: | Mobil EAL 224 |

**Evaluation**:
User reports problem free operation since 1999.

**6**

| | |
|---|---|
| **Location**: | Hanepraai Pumping Station |
| **Description**: | Underwater bearing lubric-ation/ pumping machine |
| **OEM**: | Spaans Babcock |
| **Biolubricant**: | Mobil EAL 102 |

**Evaluation:**
User reports problem free operation since 1999. Mobil EAL 102 is a lithium complex grease based on a synthetic ester derived from rapeseed oil. It is formulated to deal with operating temperat-ures ranging from –20 to 100°C, is resistant to water wash-out and has good EP characteristics. It meets the environmental and health requirements of the Clean

Lubrication Project in Gothen-burg. Besides, it is a BG1 grease according to the LLINCWA classification system. BG1 is thecategory of the most preferable greases.

## 7

**Location:**       Pijnacker Hordijk Pumping Station
**Description:**   Hydraulic oil for grid cleaner
**OEM:**            Landustrie
**Biolubricant:**  Shell Naturelle

**Evaluation:**
User reports problem free operation since 1999. Shell Naturelle
is a hydraulic oil based on
rapeseed oil. It is formulated to
deal with operating temperatures
ranging from −15 to 70°C. It
meets the environmental and
health requirements of the Clean
Lubrication Project in Gothenburg. Besides, it is a BH2
hydraulic fluid according to the
LLINCWA classification system.

## 8

**Location:**       Pijnacker Hordijk Pumping Station
**Description:**   Underwater bearing lubrication/open system
                    (3 pumping installations)
**OEM:**            Landustrie
**Biolubricant:**  Mobil EAL 102

**Evaluation:**
User reports problem free
operation since 1998. Mobil
EAL 102 has replaced the
product Mobilplex 47 (based on
mineral oil). The user's first
experience with biolubricants
was with UF 2 E, which is a
Mobil grease based on pure
vegetable oil. UF 2 E washed
away on the side of the
installation that is in contact
with water. Therefore the user
needed to pump extra amounts
of grease when the pumping

machine was not operating. The user changed over to Mobil EAL 102. Mobil EAL 102
does show these problems. It has moreover the technical advantage compared to the
mineral product Mobilplex 47. This Mobilplex 47 had the property that it solidifies at
the edges of the machine (Ca-soap). Biolubricants do not have this problem. This
results in less maintenance (and therefore also less maintenance costs). Mobil EAL 102

is a lithium complex grease based on a synthetic ester derived from rapeseed oil. It is formulated to deal with operating temperatures ranging from –20 to 100°C, is resistant to water wash-out and has good EP characteristics. It meets the environmental and health requirements of the Clean Lubrication Project in Gothenburg. Besides, it is a BG1 grease according to the LLINCWA classification system. BG1 is the category of the most preferable greases.

**9**

| | |
|---|---|
| **Location:** | Katwijk Pumping Station |
| **Description:** | Hydraulic oil for grid cleaner |
| **OEM:** | Werkspoor |
| **Biolubricant:** | Mobil EAL 224H |

**Evaluation:**
User reports problem free operation since 1999. Mobil EAL 224H is a hydraulic oil based on rapeseed oil. It is formulated to deal with operating temperatures ranging from –25 to 80°C. It meets the environmental and health requirements of the German *Blauer Engel*. Besides, it is a BH3 hydraulic fluid according to the LLINCWA classification system.

**10**
**Location:**
Katwijk Pumping Station
**Description:**
Grease for underwater bearings
**OEM:**
Werkspoor
**Biolubricant:**
Mobil EAL 102

**Evaluation:**
User reports problem free operation since 1999.

## 11
**Location:**
Katwijk Pumping Station
**Description:**
Gear oil for gearboxes
**OEM:**
Wülfel
**Biolubricant:**
Shell Omala EPB 220

**Evaluation:**
User reports problem free operation since 1999.

## 12
**Location:**      Reewijk Radenburg water treatment plant
**Description:**   Bearing lubrication of strain press
**OEM:**           SKF
**Biolubricant:**  Mobil EAL 102

**Evaluation:**
User reports problem free operation since 1996. Mobil EAL 102 is a lithium complex grease based on a synthetic ester derived from rapeseed oil. It is formulated to deal with operating temperatures ranging from –20 to 100ºC, is resistant to water wash-out and has good EP characteristics. It meets the environmental and health requirements of the Clean Lubrication Project in Gothenburg. Besides, it is a BG1 grease according to the LLINCWA classification system. BG1 is the category of the most preferable greases.

## 13
**Location:**      Spaarndammer locks (De Spaarndammer sluizen)
**Description:**   Hydraulic oil
**OEM:**           unkown
**Biolubricant:**  Ecosyn HEB 32

**Evaluation:**
User reports problem free operation since many years. Ecosyn HEB 32 is a synthetic hydraulic fluid based on mixture of esters with more than 90% biodegradability according to CEC L-33-A94.

**14**
| | |
|---|---|
| Location: | Spaarndammer bridges |
| Description: | Grease |
| OEM: | unkown |
| Biolubricant: | Castrol |

Evaluation:
User reports problem free operation since 2000.

15
| | |
|---|---|
| Location: | Oude Rijnstromen fresh water and polder mills, 5 pumping stations |
| Description: | Bearing lubrication, grease |
| OEM: | unkown |
| Biolubricant: | Vollenhoven UWS.LFB.super 2 |

**Evaluation:**
User reports problem free operation since 2000. Change took place without taking prior special measures or care.

**16**
| | |
|---|---|
| **Location:** | Hoogheemraadschap Delfland, water quality and quantity management and polder water management, 7 pumping stations |
| **Description:** | 1. Bearing lubrication |
| | 2. Gear box lubrication |
| | 3. Hydraulic oils |
| **OEM:** | unkown |
| **Biolubricant:** | 1. Castrol Biotec HM |
| | 2. Castrol Tribol Bio Top 1418/460 |
| | 3. Castrol Carelub Hes |

**Evaluation:**
User reports problem free operation since 2000. The equipment was directly filled without prior taken special measures. At the last revision no unacceptable signs of damage or wear were observed

### *Hoogheemraadschap Hollands Kroon*

Hollands Kroon is another Dutch water board where field tests with biolubricants have been conducted. It has control about the region between the *Noordzeekanaal* en *Texel*.

**17**

| | |
|---|---|
| **Location**: | Hollands Kroon,Anna Paulowna, NL |
| **Description**: | 6 grid cleaners |
| **OEM**: | Bosker & Zonen, Termunterzijl, Groningen |
| **Biolubricant**: | Mobil  EAL 224H. |

**Evaluation:**

User reports problem free operation since the summer of 2000 for 3 of the grid cleaners and March 2002 for the remaining ones except for minor operating problems at operating temperatures below 0 °C  that have been overcome.  Each grid cleaner contains around 20 l of oil and their working period is around 800 h/year with a stand period of the oil of around 1.5 – 2 years.  Owing to the fact that the biohydraulic oil functions at least equally well as the mineral oil based one in the grid cleaners, the user intends to use the same biolubricant in the hydraulic system of a reed-cutting boat owned by Hollands Kroon. The purchasing price of the biolubricant is around 20% higher than the corresponding mineral one. However, these costs make up only an insignificant part of the total budget of the department.

The user decided to switch from Mobil DTE 15 M (mineral oil based) over to the biohydraulic fluid Mobil EAL 224H after a tube breakage in 2000. The system was cleaned with a cheap biolubricant and left for one night for the remaining oil to pour out of the system. At regular intervals the oil was monitored by Mobil to check the viscosity, solidification temperature and water content. So far results indicate no unusual behavior. Mobil EAL 224H is a hydraulic oil based on rapeseed oil. It is formulated to deal with operating temperatures ranging from –25 to 80°C. It meets the environmental and health requirements of the German *Blauer Engel*. Besides, it is a BH3 hydraulic fluid according to the LLINCWA classification system.

### 7.2.2 Water management / Belgium

**18**

| | |
|---|---|
| **Location:** | Molenbeek's lock, Port of Brussels |
| **Description:** | Different joints of the locks mechanisms |
| **OEM**: | unknown |
| **Biolubricant:** | FUCHS Plantogel 2N |

**Evaluation:** Since 2001Fuchs Plantogel 2N has replaced TEXACO Multifac EP2 for the lubrication of different joints of locks mechanisms. Fuchs plantogel 2N is a Lithium / Calcium grease based on a synthetic ester. It is classified in the German Water Hazard Class 1

**19**

| | |
|---|---|
| **Location:** | Ship lift of *Ronquière* on the Charleroi-Brussels' channel[32] |
| **Description:** | Bearings |
| **OEM**: | SKF |
| **Biolubricant:** | 3 products were tested: |

- TEXACO Biostar LC EP 2 (synthetic ester with lithium/calcium soap),
- Q8 Renoir (synthetic ester with lithium complex soap)
- SKF LGGB2 (synthetic ester with lithium/calcium soap)

---

[32] The ship lift comprises two water-filled cages into which the vessels are placed. The maximum height to overcome is 67.73 meters. The castors inclined by 5 %, which results in a length of 1432 meters between the two endings. The cages move up and down the slope on sets of rollers which are continuously filled with grease and put under pressure. It is a relatively closed system, but with a hole for discharging overspills of grease. Each cage is balanced by a counterweight and is supported by 236 rollers with 70 cm in diameter, set in two rows of 59 axles. The cage and counterweight are connected by 8 cables of 55 mm diameter (x 2 cages).

**Evaluation:**
Since July and October 2002 TEXACO Biostar LC EP 2, Q8 Renoir and SKF LGGB2 have replaced ESSO Beacon EP 2 at 3x4 of the 888 rollers of the ship lift.
The lubrication with the biogreases is being done manually. Until now, problem free operation has been reported.
TEXACO Biostar LC EP 2

and Q8 Renoir meet the environmental and health requirements of the Clean Lubrication Project in Gothenburg. Besides, they are BG1 greases according to the LLINCWA classification system. BG1 is the category of the most preferable greases.


**20**

| | |
|---|---|
| **Location:** | Inclined plan of Ronquière on the Charleroi-Brussels' channel |
| **Description:** | Cables |
| **OEM:** | SKF |
| **Biolubricant:** | TEXACO Biostar Chainbar 60 |

**Evaluation:**
Since 2001 user report problem free operation.

**21**

| | |
|---|---|
| **Location:** | Ath's manual lock Canal Blaton – Ath and Dendre river |
| **Description:** | Grease lubrication of manual locks |
| **OEM:** | unknown |
| **Biolubricant:** | FUCHS Plantogel 2SRS |

**Evaluation:**
User reports problem free operation since January 2001. Fuchs Plantogel 2N is a Lithium / Calcium grease based on a synthetic ester. It is classified in the German Water Hazard Class 1.

The locks are greased manually once a year. The system is closed. Nine manual locks are involved in the test: 3 are greased with mineral grease, 3 other with a vegetable based grease without particular cleaning when applied, and 3 more with the same vegetable grease but conscientiously cleaned before the application of the new grease.

Normally, the grease removed during "drain operation" is not lost in water, but it's not completely avoidable that some grease go into water. As this waterway is mainly used for recreation, the use of biodegradable products is exceptionally important. Little by little, the old systems are replaced with Nylon parts that will not need grease anymore.

### 7.2.3 Water management / France

**22**

| | |
|---|---|
| **Location:** | Lock over the Canal des Deux Mers in Ecluse des Minimes-Toulouse |
| **Description:** | Hydraulic system of automatic lock |
| **OEM:** | CTDI hydraulique pneumatique |
| **Biolubricant:** | Bio hydro grade 32 (HAFA) – hydraulic fluid |

**Evaluation:**
User reports problem free operation since May 2001. Anal-ysis on the parameters TAN and viscosity, metal contents show a stable operation, no alteration of fluid properties. A panel indicat-es that the lock operates with biolubricants

**23**

| | |
|---|---|
| **Location:** | Canal des deux mers, (Ecluse de Négra) - Toulouse |
| **Description:** | Manual locks lubricated by grease |
| **OEM**: | unknown |
| **Biolubricant:** | IGOL (XOL GR1) - grease |
| **Since:** | July 2001 |

**Evaluation:**
User reports problem free operation since July 2001. Appreciation given according to criteria shows that the users are satisfied with the grease.

## 7.2.4 Water management / Germany

The German water way administration already started the introduction of bio lubricants more than ten years ago and therefore in general possess long term experience with bio degradable fluids. The pilot project in this field, the flood protection lock "Billwerder Deich" is therefore representative. It is a new installation rather than a substitution. The amount of 50.000 l hydraulic fluid is also quite typical. In Germany the operators of hydraulic installations are well aware of the cost effectiveness of saturated synthetic esters. They strictly use this type of fluid labelled with the Blue Angel as officially recommended by the Federal Institute of Traffic Engineering.

**24**

| | |
|---|---|
| **Location**: | Sperrwerk Billwerder Deich. |
| **Operator:** | Hamburg Ministry of economies, Amt Strom- und Hafenbau |
| **Description**: | The reconstructed Sperrwerk consists of 8 flood gates ( 2 x 34,5 m and 2 x 30 m in each line). The hydraulic driving unit consists of 1 major driving unit with 4 pumps and an additional auxiliary pump, in total 16 cylinders with 11 m stroke, 5000 m hydraulic conductions filled with 50 000 l biodegradable hydraulic fluid (saturated synthetic ester based). For maintenance purposes a mobile by pass filtering unit |

is in circulation on several plants mainly to minimise water contamination of hydraulic fluids.

**OEM:**          Herion Systemtechnik
**Start**:          August 2001
**Biolubricant**:   synthetic ester, Blauer Engel

**Evaluation:**

The routine oil trend analyses done by the lubricant supplier have been evaluated within LLINCWA and showed no alteration of fluid properties.

The Amt Strom- und Hafenbau is skilled with special know how in planning, projecting and supervision of hydraulic engineering projects, required for project management of the rebuilding of the lock Billwerder Deich as well as former projects (e.g. Estesperrwerk). The following reasons for the decision on the fluid have been mentioned:

- For Strom- und Hafenbau it is important to avoid long-term damages of the ecosystem even in case of a disaster by using ready biodegradable non toxic lubricants.
- Very good experiences with synthetic fluids in comparable plants over several years.
- Because of the particularly favourable temperature-viscosity behaviour of the fluid neither heaters nor coolers are required in contrary to mineral oil hydraulic fluids. Disclaiming warmers additionally avoids ageing of the fluid.
- The expected lifetime of synthetic esters is at minimum twice of comparable mineral oil. In plants filled with mineral oil after 2-3 years massive oil sludge was found all over the plant which required separation, cleaning of the plant and replenishment of fresh oil. After five years complete replacement was required.
- A few years ago a malfunction of a plant resulted in intense heating thereby mineral oil in one of the hydraulic cylinders deflagrated completely while the synthetic ester in another cylinder was surprisingly robust and still operates within the plant.
- Last but not least regular far reaching oil-trend analysis done by the supplier of the synthetic ester not only allows for evaluation of operativeness of the oil but additionally enables conclusions on the status of the plant itself. This supplier's service is of great value for Strom- und Hafenbau and provides information ensuring trouble free and secure operation.

### 7.3.1 Inland shipping and floating mobile equipment / The Netherlands

**25**

| | |
|---|---|
| **Location:** | ms Audri (inland marine containe vessel) |
| **Description:** | 2 open stern tubes |
| **OEM:** | De Waal, Werkendam, NL |
| **Biolubricant:** | Fina Biomerkan RS. NLGI 2/3 |

**Evaluation:**

Since April 2000 Fina Biomerkan has been used for the lubrication of the open stern tubes of Ms Audri, an inland marine container vessel from Mr Verwey, with 2 Caterpillar 3412 main engines. (508 kW/1800 rpm). The 2 open stern tube systems were in the past lubricated by mineral Calcium greases. The biogrease proved to be compatible with the mineral grease present in the system. For the change over no special treatment was necessary. There was also no change in grease consumption of 360 kg/year. No unusual behavior of the biogrease and/or unusual wear or leakage of the stern tube is observed. The technical performance of the biogrease is at least the same compared with mineral greases. The price of the biogrease Biomerkan RS is about 35% higher then the mineral grease. User and OEM are satisfied. Same experience is obtained with 4 other inland marine vessels with open stern tube systems. Cost/benefit is lower when using Fina Biomerkan RS. Fina Biomerkan RS meets the environmental and health requirements of the German *Blauer Engel,* Vamil and the requirements for the BG2 class of the LLINCWA classification system.

**26**

| | |
|---|---|
| **Location:** | Pushing Ship " Riad" 2x 3508 E Caterpillar, 765kW/1600 rpm |
| **Description:** | Closed sterntube. (Oil) Open rudder system  (Grease) |
| **OEM:** | De Waal, Werkendam, NL |
| **Biolubricant:** | Fina Biohydran TMP 100   (Oil) Fina Biomerkan RS  (Grease) |

**Evaluation:**

Since October 2000 Fina Biohydran TMP 100 (lubricating oil) and Fina Biomerkan RS (lubricating grease) are being used/tested in the new pushing ship of Mr. W. Weima.

Fina Biohydran TMP 100 is being used in the closed stern tubes. The oil content of the stern tubes is 40 liter each. After 8400 running hours (more that two years) no unusual behavior of the oil is observed. The regular analysis of oil samples shows that all the oil parameters (wear metals, oil viscosity and water content) are comparable with mineral oil. The oil-change is expected to take place after three years of operation and it will coincide with general large maintenance of the ship. Price Biohydran TMP 100 is 250% higher than mineral oil.

Fina Biomerkan RS used in the Rudder system, which is an open grease lubricated system. The grease consumption at all four rudder systems is 170 kg/year (Same grease consumption compared with mineral oil greases). No unusual behavior of the grease or the rudder system is observed. Both user and OEM are satisfied with the operation of the biolubricants. "De Waal" has fabricated both stern tube systems and the rudder systems.

Fina Biomerkan RS and Fina Biohydran TMP 100 meet the environmental and health requirements of the German *Blauer Engel*, Vamil and the requirements for the BG2 class of the LLINCWA classification system.


**27**

| | |
|---|---|
| **Location**: | Motor boat of the editorial board of the Dutch magazine for water recreation |
| **Description**: | Open stern tube |
| **OEM**: | unknown |
| **Biolubricant**: | Fuchs Plantogel 2N |

**Evaluation**:

User reports problem free operation since April 2001. Fuchs Plantogel 2N is a Lithium/Calcium grease based on a synthetic ester. It is classified in the German Water Hazard Class 1.

**28**

| | |
|---|---|
| **Location**: | Fokkelina (traditional professional sailing boat) |
| **Description**: | Open stern tube |
| **OEM**: | unknown |
| **Biolubricant**: | Fuchs Plantogel 2N |

**Evaluation:**

User reports problem free operation since June 2002. Fuchs Plantogel 2N is a Lithium / Calcium grease based on a synthetic ester. It is classified in the German Water Hazard Class 1.

**29**

| | |
|---|---|
| **Location**: | de Larus (traditional professional sailing boat) |
| **Description**: | Open stern tube |
| **OEM**: | unknown |
| **Biolubricant**: | Fuchs Plantogel 2N |
| **Since**: | July 2002 |

**Evaluation:**

User reports problem free operation since June 2002. Fuchs Plantogel 2N is a Lithium / Calcium grease based on a synthetic ester. It is classified in the German Water Hazard Class 1.

**30**

| | |
|---|---|
| **Location**: | Suction dredger (Ballast Nedam) |
| **Description**: | Open under water bearings |
| **OEM**: | unknown |
| **Biolubricant**: | Green Point BioLC 1302 |

**Evaluation:**

Since March 2002 Green Point BioLC 1302 is being used for the lubrication of the moving parts in the dredger. In dredging the sand sediment, sand grains enter into the moving parts, which is prevented by pumping grease out of the system. It is a semi-open grease lubricated systems. A mineral based lubricant was used but owing to the company's environmental image a plan was developed to prevent the mineral based lubricant from entering the water. As a first attempt rubber rings were used to seal off the system. This was not

successful. Therefore the mineral oil has been substituted by a biolubricant, Green Point BioLC 1302 from the company of Van Meeuwen, Weesp. The quantity of grease used is around 2000 kg/year.

Before changing the grease, both the old and new grease were mixed to prevent any side reactions. No change was observed. No problems were reported after the change over to the biolubricant. Green Point BioLC 1302 is a lithium/calcium grease based on synthetic ester. It is formulated to operate in the temperature range -30 to 120°C. The grease has good EP properties.

### 7.3.2 Inland shipping and floating mobile equipment / Belgium

**31**

| | |
|---|---|
| **Location:** | Libertas (school boat used for training of French-speaking skipper students) |
| **Description:** | Closed stern tube automatically greased |
| **OEM:** | unknown |
| **Biolubricant:** | FINA BioMerkan RS |
| **Since:** | May 2001 |

**Evaluation:**

After the substitution of stern tube grease on the *Libertas* ship, some greasy black spots appeared on water. After some weeks it disappeared and then showed up again. A specialist from TotalFinaElf inspected to the equipment and took samples. It seemed that when replacing the grease in the supplying bottle feeding the tube, the old grease remained inside the tube. The grease brought to the tube arrived near the extremity at the water side. Supposing that the two greases were not compatible, tests were done in laboratory, confirming that the mineral one separated into oil and soap, and could create the black spots on water.

It was decided to clean the contents of the tube, to make sure that the all existing grease in the stern tube was removed. Actual substitution experience (in the Netherlands) where FINA products were substituted by FINA BioMerkan RS no compatibility problems were observed.

After thorough inspection, putting the ship in a dry dock, it appeared that a leakage of black oil was coming from the chamber above the machine room, just close to the propeller. This turned out to be the substance that created the black spots on water, instead of the new grease in the stern tube!  The lesson learned from this substitution experience is to be extremely careful in drawing unbalanced conclusions and not automatically blaming the bioproduct.

During the reset of the axis in the shipyard, a precise measurement of the axle was made, to be able to judge wear in the coming three years time. The result of the experiment is a good operating Biomerkan  biogrease lubricated *Libertas.* The skipper is fully satisfied, both for the water tightness and the lubrication properties of the grease. He decided to grease the new  "stem" motor's stern tube with this biogrease as well.

### 7.3.3 Inland shipping and floating mobile equipment / France

**32**

| | |
|---|---|
| **Location:** | Canal des Deux Mers (between Narbonne and Montauban) |
| **Description:** | Hydraulic system of a dredger working for the maintenance of the canal des Deux Mers |
| **OEM:** | Watermaster |
| **Biolubricant:** | Eco hyd S plus (FUCHS LUBRITECH) hydraulic fluid |
| **Since:** | May 2001 |

**Evaluation:**

User reports problem free operation since May 2001. Analyses of fluid samples every 250 h (TAN, viscosity, metal contents). Based on these analyses it the lubricant is shown to be stable, there is no alteration of fluid properties and no typical wear.

### 7.3.4 Inland shipping and floating mobile equipment / Germany

**33**

**Location:**       Working pontoon "Annegret" on the river Elbe
**Operator:**       Bilfinger & Berger, MT Nord
**Description:**    Hydraulic system and various grease applications encountered on the
                    pontoon
**OEM:**            J.L. Muns B.V
**Biolubricant:**   Panolin HLP Synth 46 – hydraulic oil DEA Dolon EP2 f- grease
**Since:**          March 200

**Evaluation:**

The pontoon is a working pontoon 40m long, 20m wide, 3m high with legs measuring 40m. The 6 hydraulic mooring winches with a total force of 9 t each and the stilt- system are driven by the same hydro-units each 360 kW. Below deck there is also the three-phase diesel generator and the

diesel tank (60 m$^3$), a hydraulic tank with a 5600 l capacity. Shunt current filtration ensures a low water content in the hydraulic oil and also ensures long-term use.

Approx 8000 l of biodegradable hydraulic oil (synthetic ester) is being used for the hydraulic components (pontoon legs and winding). Biodegradable grease is being used for the legs and trunk.

Costs of the lubricants in relation to the total construction costs are 5.500 thousand DM. The hydraulic oil cost 58 thousand DM, more than double that of a comparable mineral oil. There are then the costs of the filter unit of 11 thousand DM (1.3% of the construction costs). The biodegradable grease is a consumable. As an indication of the running costs, a figure of 7.2 thousand million DM was given for the pontoon's consumption during the first 3 months of safety work carried out on the Elbe-tunnel. The additional costs for biolubricants are therefore considerable. They constitute protective measures for ensuring the safety of the plant and minimising the damage caused. There are other measures such as safety valves to reduce the possibility of a leakage or accident in the hydraulic system.

A further cost factor is the fail-safe and operating safety of the plant. The shunt current filter installation (oil maintenance) and the bi-annual oil analysis helps to ensure that the plant operates without malfunctioning. They ensure durability and provide early warning of a possible malfunction or failure of the system.

B & B expects an utilisation rate of 60 to 80% for the new pontoon as for its predecessor. It has been in operation since April 2001, with a two month break for maintenance work, and since November 2001 has been operating on a long-term basis again in Bremerhaven. Only a fail-safe plant can guarantee a high utilisation rate and the exercising of tasks with the lowest possible risks. In brief, it can be said that, for Bilginger and Berger, the additional costs for biodegradable lubricants are offset by their benefits. The benefits are:

- Fulfilling of legal duties of the plant operator, namely to avoid dangerous events and to minimise the possible extent of damage.
- Plant safety
- Reduced maintenance costs through longer service life.
- Fail-safety/operating safety
- Planning stability

**34**
**Location:**
Two bridge inspection boats of the German Waterways Administration (West and East)
**Description:**
Hydraulic system
**OEM:**
Bröhl (hydraulic legs) Ruthmann (rising gate)
**Biolubricant:**
Panolin HLP Synth 46 for hydrailic legs
Aral Forbex E22 for rising gate

**Evaluation:**
Problem free operation since September 2001. Oil trend analysis is still ongoing.

**35**

**Location:**  Survey, wreck-search and research vessel DENEB belonging to the
Federal Maritime and Hydrographic Agency of Germany

**Description:**  The following components are operated with biolubricants:
1) Simplex Stern tube,
2) Schottel hydraulic system,
3) Hatlapa winch hydraulics,
4) Atlas crane hydraulics

**OEM:**  unknown

**Biolubricant:**  Shell fluid BD 46 for application 1
Shell Naturelle HF-E 46 for application 2, 3 and 4

**Evaluation:**

The demonstration project has been chosen to evaluate long-term experiences. The staff on board confirmed failure-free operation of the hydraulic systems and the stern tube. In daily operation systems with biolubricants are treated the same as any other system without special care. Within LLINCWA it

was tried to make earlier oil analysis results available. However, we failed because obviously neither the lubricant supplier nor the ship inspector or the chief retained the protocols of oil analyses.

Recently the stern tube has been changed over to mineral oil. The ship inspector reported, that he had no other choice, because Shell fluid BD 46 is no longer being produced. The lubricants supplier stopped production of this product because of too low demand.

**36**

**Location:**    Fishery protection boat "Seeadler" belonging to the Federal Institute
of Fishery

**Description:**    The following components are operated with biolubricants:

1. Hatch cover,
2. Davit
3. Provisions crane
4. Warping capstan
5. Anchor winch
6. Propulsion rudder
7. Valve remote control,
8. Warping and working winches

**OEM:**    unknown

**Biolubricant:**    Aral Vitam EHF 46 for application 1 and 8; Aral Vitam EHF 22 for
application 2, 5 and 7, Aral Degol BAB 150 for application 3, 4 and 6

**Evaluation:**

No special evaluation has been performed. This boat perfectly shows the variety of on-board applications that can be substituted by bio lubricants.

The staff on board confirmed failure-free operation of the hydraulic systems and the stern tube. In daily operation systems with bio lubricants are treated the same as any other system without special care.

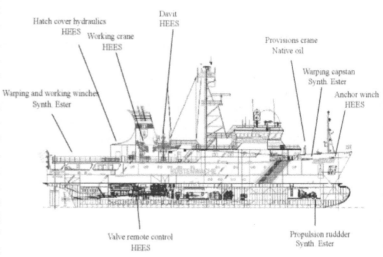

### 7.4.1 Hydroelectric power / Austria

**37**

| | |
|---|---|
| **Location:** | Hydroelectric power station "Langkampfen" on the river Inn, Austria, operated by TIWAG |
| **Description:** | The hydroelectric power station "Langkampfen" of the TIWAG runs various applications with biolubricants |
| **OEM:** | unknown |
| **Biolubricant:** | Panolin Turwanda Synth. 68 |
| | Panolin HLP Synth |
| | Panolin HLP Synth 32 |
| **Since:** | 15 years |

**Evaluation:**

The hydrolelectric power station "Langkampfen" of the TIWAG runs the following applications with biolubricants:

Generator and turbine bearing, turbine controller, Weir fields, Station crane, and the rake cleaning machine.

As an overall conclusion it was stated that mineral oil based lubricants were substituted successfully.

An extensive description of the extensively documented substitution of this pilot project is given in chapter 8.

### 7.4.2   Hydroelectric power/France

**38**

| | |
|---|---|
| **Location:** | Garonne river, Toulouse, France (RMET : régie municipale d'électricité de Toulouse) |
| **Description:** | lubrication of racks with grease |
| **OEM:** | unknown |
| **Biolubricant:** | grease BIOMERKAN RS (TotalFinaElf) |
| **Since:** | July 2001 |

**Evaluation:**

The flood gates are under water half time and are exposed to sun rays (extreme conditions). Grease was judged not enough resistant to water. The presence of white residue was noticed after sun exposition. The frequency of application was multiplied by three, which takes too much time for users. Tests with a new grease are in progress.

### 7.4.3   Hydroelectric power/Spain

**39**

| | |
|---|---|
| **Location**: | Hydroelectric turbine |
| **Description**: | Centralized grease lubrication for the low turbine bearing |
| **OEM**: | GFAlsthom equipment |
| **Biolubricant**: | BIOVERKOL CR-2  (Verkol) |
| **Since**: | February 2002 |

**Evaluation:**
In the past they used a multipurpose lithium EP-1 grease. With the biogrease, the pressure in the centralized system is slightly higher because it is a NLGI 2 grade grease

## 7.5. MISCELLANEOUS

**40**

| | |
|---|---|
| **Location**: | Armstrong Nederland, Hoogezand, The Netherlands |
| **Description**: | Open chain lubrication (open system) |
| **OEM**: | Tailor made machinery built by Armstrong Nederland |
| **Biolubricant**: | Mobil Setac EAL 68 (Chain-saw oil) |
| **Since**: | February 2000 |

**Evaluation:**
A part of the production process consists of a drying machine of ~70 m high with several levels, driven by a chain. The chain was sprayed by a mineral oil that owing to the higher temperature of the oven (120 °C) contained a substantial high amount of high temperature and anti-corrosion dopes. However, owing to the spraying process the mineral oil fell down and caused a greasy floor. The oil also entered into their process water, with a stand period of around 30 days before it is discharged into the public sewage system. Remnants of the oil in the process water increased slowly. Both the safety aspects and the increasing impurity of the process water were the main drivers to change the lubricating grease.

A chain-oil biolubricant was selected that was rather viscous and was applied manually once a week onto the chain.

Owing to the change of application a reduction in quantities was obtained of around 75%. A second aspect was that the price of the biolubricant chain-oil was around half of the mineral oil one! Oil measurements in the process water before discharge indicated that the oil had disappeared within the 30 days period. Owing to the fact that the grease/oil was applied manually, the floor remains clean.

The main disadvantage is that the quality of the chain cannot be controled anymore by simple visual inspection. Therefore a preventive check-up was built in the production process whereby the chain is removed once a year for inspection. Within the two years period no extra corrosion or degradation was observed.

**41**

| | |
|---|---|
| **Location:** | Canal des Deux Mers, Toulouse, France |
| **Description:** | Excavator working for the maintenance of the bank (Canal des Deux Mers) |
| **OEM:** | Akerman |
| **Biolubricant:** | hydraulic fluid   Biohydran SE 46 (TotalFinaElf) |
| **Since:** | January 2001 |

**Evaluation:**
Analyses of fluid samples were taken every 250h (TAN, viscosity, metal contents). The parameters show a stable behaviour of the lubricant, there is no alteration of fluid properties and no typical wear. The seal that has been changed has revealed mechanical wear without chemical alterations.

**42**

| | |
|---|---|
| **Location:** | NORMA company, Spain |
| **Description:** | Bending operation |
| **OEM:** | PICOT |
| **Biolubricant:** | DECOLUB Ecopipe, BRUGAROLAS |
| **Since:** | January 2001 |

**Evaluation:**
The biolubricant performed 14 month without any problems. It was concluded that the oil performs very good

# Chapter 8

# Practical tips for the substitution of lubricants

This chapter provides engineering personnel with general design guidance to select, specify, inspect, and approve biolubricants for their equipment. Changing from conventional mineral oil based lubricants to biolubricants is not always simply a matter of one-to-one substitution. Furthermore, biolubricants are not necessarily equal in performance to one another. In order to make the switch over a success the following properties of biolubricants should be given thorough consideration.

It generally is recommended that the OEM be consulted prior to the initial purchase of any biolubricants. Moreover, an operator considering changing over to biolubricants should make sure that the supplier is aware of the pros and cons of biolubricants.

When selecting a biolubricant the following rule of thumb holds. Vegetable oil based lubricants are typically used in open and low-tech applications. For more sophisticated applications you may have to chose esters and, in some cases, polyglycols.

### Oxidation and thermal stability

Oils with low values of oxidation stability will oxidize rapidly at elevated temperatures. When oil oxidizes it will produce acid and sludge. Sludge may settle in critical areas of the equipment and interfere with the lubrication and cooling functions of the fluid. The oxidized oil will also corrode the equipment. To check the oxidation stability of biolubricants one should use the so called dry-TOST test, a laboratory ageing test in which the age of degradation of fluid is determined by measuring the increase of acid number (TAN). The test can be performed according ASTM D 943, except that no water is added to the fluid. Another method to evaluate oxidation stability is the RBOT ASTM D2272.

Vegetable oils in general, do not poses good resistance to oxidation, but high oleic vegetable oils do have acceptable oxidation stability for a wide range of applications. Polyglycols and synthetic esters are in general more stable than most vegetable oils. Some properly formulated synthetic esters show even superior oxidative stability than mineral oil based lubricants.

Biolubricants have bad stability at high temperatures. But they are suitable for all applications when temperature is below 70 °C, and Esters of vegetable oils are suitable below 120 °C. The temperature stability of vegetable oils has been improved through oleic basis and additivation. Time dependence low temperature properties should be also checked.

*Low temperature behaviour*

The low temperature fluidity of vegetable oil based fluids is poor compared to other biolubricants. When oil solidifies, its performance is greatly compromised. However, the low temperature behaviour of vegetable oil based fluids may be acceptable for most applications. For determining the low temperature behaviour of fluids, the pour point normally is measured according to ASTM D97. However, the method described in the VDMA specification 24568 is recommended as more appropriate for investigation of the low temperature behaviour of esters.

*Hydrolytic behaviour*

This is a point of concern in particular when considering the use of vegetable oils and to a lesser extend when considering the use of synthetic esters. The hydrolytic stability of commercially available biolubricants is enhanced by means of additivation. The hydrolytic stability is normally measured by ASTM 2619 (Beverage-Bottle-Test).

*Compatibility with system components*

Biolubricants may not be compatible with some paints, finishes, as well as some seal materials and metals. This is a point of concern especially in the case of polyglycols and to a lesser extend in the case of synthetic esters. Extensive practical experience with vegetable oil based lubricants has yielded relatively few problems with seals and paints. Moreover, vegetable oil based lubricants are good compatible with steel and copper alloy and provide excellent rust protection. However, the supplier must be always consulted for specific compatibility data for each material encountered in the application. Some general rules are depicted here bellow:

- Standard paints are incompatible with most biolubricants. It is recommended that epoxy resin paints be used with biolubricants.
- Polyurethane should not be used with biolubricants. Instead, Viton® and Bunna N (low to medium nitril) are mostly recommended with biolubricants.
- Polyglycols are not recommended in installations comprising aluminum.
- Biolubricants classified as 1a and 1b according ASTM D 130 are compatible with copper alloys.

*Filters compatibility*

Special filter elements are not required in the case of vegetable oils and synthetic esters. Paper filters may need to be replaced with glass-fibres or metal-mesh filters when changing over to polyglycols. When changing over to biolubricants filters should be checked after some hours of operation, as biolubricants tend to remove mineral-oil deposits from the system and carry these to the filters. Filter-clogging indicators should be carefully monitored.

*Foaming*

The tendency of oils to foam can be a serious problem in lubricating and hydraulic system. The lubrication and hydraulic properties of oils are greatly impeded by excessive foaming. Foaming characteristics can be evaluated by using ASTM 892-89. Laboratory tests have show that most formulated biolubricants do not have foaming problems.

### Residual mineral oil

When changing over to biolubricants, the system should be preferably drained of the mineral oil and, if possible, flushed. In the case of polyglycols 1% of mineral oil residue is acceptable. In the case of vegetable oils and esters a maximum oil residue of up to 2% is allowed.

### Costs

Biolubricants are more expensive than mineral oil based lubricants. The purchase price difference is significant. In 1999 the German Federal Ministry of Agriculture investigated price ranges of different lubricants, and identified the following price ranges.

Table 8.1   Price range of lubricants and hydraulic fluids (1999)

| Application | € /100 Liter  -   € /100 kg | | | |
|---|---|---|---|---|
| | Native oil | Synthetic Ester | Glycol | Mineral oil |
| Loss lubrication | 125 - 400 | ca. 325 | x | 75 - 500 |
| Chain saw oil | 125 - 1750 | x | x | 85 - 130 |
| Lubricants | 150 - 225 | ca. 700 | ca. 400 | 145 - 400 |
| Hydraulic fluids | 100 - 300 | 200 - 600 | ca. 425 | 43 - 250 |
| Gear oil | 350 - 700 | 350 - 800 | 375 - 750 | 90 - 500 |
| Grease | 240 - 300 | 550 - 2750 | x | 125 -550 |
| Motor oil | x | 300 - 1100 | x | 125 - 700 |

**x:       minor importance / does not exist / no information available; Source: Report on rapidly biodegradable lubricants and hydraulic fluids of the German federal Ministry of Agriculture, 1999**

By selecting a biolubricant that requires the least adaptation of the materials in the lubricated system the costs can be lessened.

By monitoring oil during use, it is possible to increase oil duration changing the oil when it lifetime is finish, not after a preventive fixed deadline. In this way, it is possible to reduce generation of contaminant residues and maintenance cost. The next table gives an overview of the tests that can be performed for oil monitoring purposes.

Table 8.2  Physical-chemical analysis of used oils

| Parameter | Measurement | Protocol |
|---|---|---|
| Physical properties | Viscosity | ASTM D 445 |
| | | ASTM D 2270 |
| Oxidation | TAN, Acid number | ASTM D 664 |
| Wear particles, additivation, | ICP, Plasma | ASTM D 5185 |
| Contamination | PQ, Wear index | PE-5024-AI |
| | | |
| Friction and wear tests of new biodegradable oils/mineral oils | | |
| Adhesion resistance | Coef. Friction, wear | P11.10/001 |
| Abrasion resistance | Coef. Friction, wear | P11.10/003 |

Finally, the total costs of using biolubricants are influenced by the following factors often in the direction that result in a more competitive position for biolubricants:
- tool life which is at least equal and often higher
- less evaporative loss (vegetable oils are about 5 times better)
- less oil loss due to greater adhesion to metal surfaces
- cost savings on maintenance, man power, storage and (possibly) disposal costs
- lower fines and lower clean-up costs in case of accidental spills
- improved environmental image
- lower environmental burden.

## GENERAL EXAMPLE OF THE TIWAG AUSTRIA SUBSTITUTION

(chapter 7, example 29)

The hydroelectric power plant TIWAG in Austria has already a long lasting experience with substitution of mineral oil based lubricants by biolubricants and is one of the users that monitored this substitution process extensively. It is therefore that the TIWAG case is presented as success story, but also as an example of the many aspects that have to be taken into account in substitution.

**Austria**

*Hydroelectric power station "Langkampfen"*

Fig. 8.1   Hydroelectric power station "Langkampfen" on the river Inn

Although the bio lubricants in this installation were not introduced during LLINCWA, but earlier, the project was chosen as a demonstration project to evaluate some long-term experience.

**Location:** Langkampfen on the river Inn.

**Operator:** The TIWAG Inc. based in Innsbruck and founded in 1924 by the state of Tyrol, is a prosperous internationally-active venture. With 9 large and 34 smaller power stations, a total capacity of approximately 1500 MW producing 3000 GWh, the TIWAG provides the major part of Tyrol with energy. The company and its employees are aware of their responsibility and attach great importance to the environmentally sound construction and operation of power plants because e.g. emissions of hydraulic fluids in the alpine terrain would cause irreparable long term damages to the sensible ecosystem. Since more than 15 years environmentally sound fluids and greases are used in TIWAG power stations.

**Description:** The hydrolelectric power station "Langkampfen" of the TIWAG runs the following applications with biolubricants
1.  Machine 1 and 2, Generator and turbine bearing
2.  Machine 1 and 2, turbine controller
3.  Weir fields 1, 2 and 3
4.  Station crane
5.  Rake cleaning machine

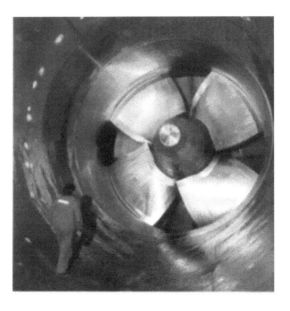

Fig. 8.2 Turbine wheel in Langkampfen

**Biolubricants:**    Panolin Turwanda Synth. 68 for application 1 and 2 (21600 litre in total)
Panolin HLP Synth 15 for application 3 (9000 litre in total)
Panolin HLP Synth 32 for application 4 and 5 (350 litre in total)

**Ecolabel:**    Blauer Engel / Panolin HLP Synth 15 and Panolin HLP Synth 32 meet also the environmental requirements according to Swedish Standard SS 15 54 34

**Evaluation:**

Aspects that identified by TIWAG to play an important role in substitution:
- Legal requirements and requirements by public authorities
- The company's public image
- Ecological requirements
- Economical requirements
- Technical requirements

**The legal situation:**

Actually Austria lacks legislation that does oblige operators to use biodegradable lubricants. Nevertheless national standards for biodegradable lubricants do exist. The EU Water directive 2000/60/EU came into force in 2000 and will be adopted into national law in 2003, while its practical implementation will be enforced since 2015. For the implementation of this directive the federal ministry of Austria already installed a working group.

**Company's public image and ecological requirements:**

As there are hardly any regulations by Austrian authorities, the choice of system fluids is lead by the acceptance of the company's responsibility and public image. To build up hydroelectric power plants encroachments to nature are considered inevitable. All considerations and measures are based on different national and international standards, mainly the WGK, the VCI-concept and the VwVwS-concept (valid for the storage of water hazardous goods).

**Economic requirements:**

For TIWAG additional costs are not an argument! Additional costs of filling a hydraulic installation of large plants with more than 5000 liter with environmentally sound fluids compared to mineral oil does account only for 0,1 to 1 per mill of total investments for the machine construction. In plants with up to approximately 2000 liter fluid the generator and turbine supplier do not charge any additional cost for filling with environmentally compatible fluids. Therefore, additional cost for environmentally sound system fluids are negligible and a bio oil discharge on the other hand would cause much less consequential loss compared to mineral oil.

**Technical requirements:**

In addition to the normal technical requirements there are a number of requirements necessary to make a product suitable for everyday work. Some of these are:

- Problems and questions with fluid handling:
  - Is a protecting closing required for the filling of hydraulic installations?
  - How does it smell, are there health consequences by skin contact?
  - How and with what should spilled fluid be wiped up?
  - Are floors, concrete coatings and paintings dissolved?
  - Is the fluid flammable or explosive?

- Problems and questions with the filling of lubricating systems:
  - Compatibility with residual assembly pastes, lubricants, cleaning agents, waxen protection coatings for transport, etc.?
  - Compatibility within the hydraulic circulation with coatings, electric wires, non-ferrous metals, plastics, etc.?
  - What about swelling and dissolving of seals, is it possible to use standard materials like NBR (nitrile butadiene-rubber) or is FPM (fluorine rubber) required? The sealing materials are in everyday use indistinguishable, therefore one can assume that there are a number of incompatible seals and for these reasons dysfunction and leakage may have to be accounted for?

- Problems and questions following lubricant emissions.
  - What is the specific gravity of the fluid, does it drift at the surface of the discharge system or sink to the bottom? How have oil separators to be built? Are spills visible on the surface water? What about miscibility with water? Will concrete coatings be damaged?

- Problems with greases in open systems:
  - In open system some loss of greases is accepted in order to guarantee system lubrication. The lubricating grease has a number of functions as e.g. reduction of oscillations, anticorrosion, sealing up the system, carriage of residuals, etc.

- For the lubricating of technically simple bearings with wide tolerance, as e.g. the wheels of rollaway contractors, an acceptable alternative to lime soap (Mobil AA2) could not be found. Using other products results in an increase of the amount of lubricants by a factor 2 to 5, which was not considered ecologically feasible. It is assumed that the higher viscosity of the base fluid and its saponification results in the good adhesiveness.

**Problem zones of an oil discharge within the plant**

During LLINCWA an oil discharge within the power plant Langkampfen happened the 31.1.2001 after 7810 hours of operation, shows exemplary that the failure of a small sealing results in considerable assembly cost and business interruption of 1 week. The shrinkage of the 8 mm square and 6700 mm length O-circlet seal at the generator bearing cover resulted in an oil discharge of approximately 100 liter into the generator.

The TIWAG undertook a detailed damage analysis in cooperation with the supplier of the turbine and generator as well as the supplier of the hydraulic fluid. During the analysis a number of seals were checked, but they were all ok. The ascertainment of damage, the cleaning of the generator and exchange of the seal resulted in a business interruption of 1 week and costs of approximately €100.000 apart from the capacity breakdown.

The supplier of the turbine and generator and of the hydraulic fluid both independently investigated the failure and concluded the following:

Cross-linking of the used NBR material and the softener did not comply with common standards and would have been shrunken equally with mineral oil. With NBR seals of well-known suppliers such problems up to now never occurred.

## Problems of oil discharges into the environment

An emission of system fluids into the discharge system is a quite frequent and awkward event of fault. Fluid emissions most likely happen during start-up. During this phase of operation some equipment is not yet operational which may result in technical failures, on the other hand human failures, like faulty operations or erection defects quite often contribute to oil discharges.

In old plants the biggest problem is the floor of the power station with its numerous fretwork, shafts and drains. In many cases the configuration as well as ground level is so unfavourable that a hermetic sealing of the building is impossible.

### Measures in case of fault

If problems, events of failure and damages occur the following measures have to be taken as soon as possible:
- Take urgent measures according to the alert program to minimise the extend of the damage
- Inform all parties whom it may concern about the incident (authorities, suppliers, insurance company, ...)
- Record the damage in detail
- Investigate start and progress of the damage
- Carry out detailed fault analysis
- Elaborate a catalogue of measures
- Realise necessary measures to ensure secure operation
- Realise customer satisfaction by adequate measures

If all parties involved contribute with their knowledge and competencies, one can make sure, that such events remain exceptions.

# Chapter 9

# Evaluation LLINCWA approach

Looking back at three years experiences with LLINCWA, many different successes and failures can be distinguished. A number of key points can be identified in the technology transfer approach chosen by the LLINCWA team that explain these successes and failures . These key points can be headed under the following items:

- LLINCWA's message: this concerns the specific argument that was put across by the LLINCWA team
- The LLINCWA team: the composition of the LLINCWA team reflects a specific choice of partners
- LLINCWA's coalitions for support: LLINCWA has allied with specific market players and societal actors in specific ways
- LLINCWA's project design: the selection of goals, sub-goals and means to achieve them represents a specific theory on how to realise LLINCWA's mission

In this chapter we reflect on the specific choices made concerning these four key points, and account for the extent to which they have furthered or hindered LLINCWA's effectiveness. Below we address the four key points successively, and discuss the consequent successes and failures of the choices made.

### LLINCWA's message

The LLINCWA message can be summarised as follows: "The use of biolubs is technically and operationally feasible, economically viable (and often profitable) and most of all: environmentally benign and necessary. Therefore substitution is called for."

LLINCWA had no problems to formulate this message and to find acceptance amongst policy makers and the general public. It is a noble activity to fight for the improvement of the fresh water quality, the reduction of diffuse water pollution and the reduced use of hazardous chemical products. No one seriously objects to that. It is even included in national and international legislation and it is formalised in general goals for industrial activities denominated as *responsible care* and *product stewardship*. This holds for as far as environmental goals are concerned. Interference in production, in established working methods and ingrained product-handling procedures is, however, a completely different matter. Here we found that the LLINCWA message encountered less self-evident acceptance, if not, sometimes, outright resistance. LLINCWA's activities appeared urgently needed, but surprisingly enough, as a response to our request for substitution the first reactions experienced were strongly dualistic: a critical acknowledgement for the environmental goals and scepticism on the performance of bio lubricants, disbelief and a cry for guarantee on technical aspects; this resulting in discussions in which the aquatic burden of mineral oils and the need for interfering initiatives were disputed.

To realise the apparently simple (environmental) improvement (substitution of the toxic, non-biodegradable lubricant) little governmental or other stimulating activity was discovered so far. It is true that mineral oils are being identified as an important source of diffuse aquatic pollution but due to other existing toxic pollutants no priority is formulated to fight the mineral oil source, neither in national nor in international legislation. This with an exception for environmentally sensitive areas where our general environmental feeling can drown out low industrial activity and realise a biodegradable alternative. As soon as smaller or larger interests start playing a role, reserved reactions do predominate towards to the principle of substitution, the source oriented approach.

LLINCWA's lubricants-user approach to increase the awareness, to show the environmental harm and to suggest substituting the polluting product can be characterised as a conflict approach. After all, LLINCWA's message that the "traditional" lubricant choice needs revision can be interpreted as a critic on actual operational management. The company is using an environmentally unfriendly product and because alternatives are available it may feel accused for irresponsible operating. The plea for *substitution* might be more acceptable if it were addressed as an *optimisation* of the process, thus avoiding the conflicting approach. Optimisation might mean implementing some simple adaptations leaving the operational process more or less unchanged, while substitution aims for a more radical change, sometimes even requiring changes in processes, equipment, brands of products and suppliers.

For the lubricant supplier another conflicting situation does appear. None of the larger lubricant (multinational) companies, and only a minor part of the smaller niche market players, have formulated bio lubricants as their spearhead trade. The LLINCWA attempt to form a coalition with lubricant suppliers in promoting bio lubricants means that suppliers themselves are asked to become their own competitors. For some larger companies this might lead to a problem for only a few of their salesmen. Some multinationals have established a small unit within their own company to promote bio lubricants, resulting in the fact that these have to compete internally and externally with their own larger units selling "traditional" lubricants. They will find their colleagues from their own companies as competitors. For environmental reasons that are often of secondary importance in their day-to-day lubricant supply business, salespeople are supposed to promote the substitution of their technically well-performing products; a substitution with less well-known and generally more expensive bio lubricants. Even if one takes into account the positive approach of most suppliers, for many of them this seems a bridge to far. They are trying to find (even ecological) arguments to protect their traditional lubricants against the attack of bio lubricants.

And so, notwithstanding its plausibility at first sight, in practice we faced a broad range of – more and less valid – arguments that were more or less openly aiming to weaken the impact of the LLINCWA message. These arguments concerned for instance the growing importance of – supposedly even more environmentally benign – water lubrication. Also, substitution was deemed unnecessary because of the use of closed systems and of the proper ways in which the bilge water is disposed of. Additional counterarguments rested on differences in the lifespan, in sparse use and in recycling methods, which would supposedly alter the environmental balance between the different lubricant types. Different testing methods were pointed at, supposedly leading to different conclusions. In all these cases, work to assess these arguments and incorporate them into the LLINCWA message cost a lot of effort – often to the effect that the LLINCWA message remained largely unchanged.

In conclusion we can say that the somewhat antagonistic nature of the LLINCWA message helped the project in that it allowed for clear positions and communications, helped to make a stir and evoke action and reaction. At the same time, however, it roused opposition that started to generate confusing or opposing messages. As such, the LLINCWA message served most often as the starting point for debate, though not seldom also as the final conclusion of the debate.

**The LLINCWA team**

A typical characteristic of the LLINCWA team was that a large proportion of the partners was not firmly rooted in the lubricant sector. Most partners had their background in research and consultancy: IVAM-CR, QA+, Valbiom, ISSUS, INPT. They co-operated with some other partners whose daily practice does indeed revolve around lubricants: the producers/suppliers Fuchs and TotalFinaElf, the tribological institute Tekniker and the user Hoogheemraadschap Rijnland.

Particularly for those groups involved in LLINCWA that do not have a background in the lubricant business, the three-year LLINCWA activities meant an intense effort to unravel the complex lubricant market structure, to identify relevant actors, stakeholders, involved parties and governmental authorities. They range from farmers growing the vegetable oil seeds to the heavy metal industry building ships, from chemical additives manufacturing industry to small niche lubricant manufacturers, from the environmental movement along the Rhine commission to the European parliament and from inland shippers along the water management boards to the recreational sector. But also for those whose first business is in lubricants, the broad scope of LLINCWA brought along a number of new perspectives. The requested skills and needed evidence were impressive: tribological skills, chemical skills, environmental and occupational health knowledge, mechanical engineering knowledge, marketing expertise, but also the ability to organise workshops, to organise conferences, to make effective PR etc.

Nobody can be accused to unite all these skills within one person, and a key point in these type of technology transfer activities therefore seems to be the formation of a tightly coherent project team with a powerful national and a broad international reach. The team must combine relevant technical and environmental expertise and must be able to formulate its goals in an appealing way: the team must have charisma.

As an organisation of teams with differing backgrounds, from different EU-counties and generally not acquainted with each other preceding the actual start of the project much attention has to be paid to the formation of a tight, solidary team with a common understanding of the project goals avoiding language or cultural problems. A team that feels the spirit to go for it. This fighting spirit seems an indispensable tool to overcome the outlined conflict-oriented message (see before). "If we can't reach our goals this way around, we will take the other".

In this respect one might wonder if the chosen LLINCWA team and the chosen coalitions have been an optimal choice. If we look at the LLINCWA team somewhat closer, we can distinguish three different actor types:

- *Knowledge providers / environmental experts / lobbying experts.* In this group will fall the academic, technical and consultancy organisations IVAM-CR, QA+, ISSUS, Valonal, Tekniker and INPT
- *Suppliers.* TotalFinaElf and Fuchs Lubritech
- *Users.* Hoogheemraadschap van Rijnland

Very clear one can distinguish oppositional interests between these groups. The first group does operate within the frame of its formulated environmental goals, including several topics of scientific interest, but does not have any commercial interest in the lubricant market. Their task is to take care for the actual independent technology transfer, organise the supporting scientific evidence (research and setting up of pilot projects) and dissemination of the message. Nevertheless the national LLINCWA partners have been selected based on their scientific skills and they were not necessarily skilled in performing the technology transfer job.

An important common characteristic of this group (perhaps with the exception of Tekniker) is that they have no vested interests and no established contacts with the lubricant sector. They are more or less free to take any position they want, to select temporary allies and opponents, and to choose the course of action they see fit.

The latter characteristic does not necessarily hold for the other two groups (and, as said before, possibly neither for Tekniker). The second group, consisting of suppliers, has a strong interest in selling lubricants. It is their business and although they committed themselves strongly to the goals of LLINCWA, their LLINCWA-role cannot be seen separated from their trading business. Their orientation is strongly dominated by the market demands for lubricants and they are, of course, always ready to adapt their supply to the demands of OEMs and the explicitly formulated users demands. The salesman has to assure his daily sales.

Nevertheless, to reach the LLINCWA-goals, LLINCWA might need to develop branch-conflicting initiatives to force a necessary breakthrough, which might not be easy (or even impossible) to be accepted by interested parties like suppliers.

The interest of the third group of users is dualistic. They combine their generally formulated policy to take care for the water quality with their need to operate reliable and (financially) efficient. The motivated technician or responsible person for environmental affairs might come in conflict within their company with the purchasing department or management that has to take care for the available budgets.

The conflict on the establishment of the ranking system is a good example in this respect. The establishment of a transparent system for the recognition of acceptable bio lubricants turned out to be a hornets' nest. Four groups of biolubs were defined, based on unambiguous set of scientific parameters, to rank and identify the acceptability (on environmental and health hazard terms) of lubricants. In first instance performance parameters were excluded from this system, to emphasize the environmental approach. Nevertheless the suppliers group with different arguments objected the establishment of a ranking system. Most important in this respect was the argument that they did not like a ranking. Due to technical demands they are forced to use specific additives that might result in a ranking of their products in group four, the lowest of the system (but nevertheless still an acceptable bio lubricant). The name ranking system was skipped and the name classification system was introduced. The compromise was found in a pragmatic approach that combines scientific analysis and the need to bring into line existing national policies on bio lubricants. LLINCWA formulated a *set of minimum requirements* (for lubricants to be designed as *bio lubricant*), but yet less rigorous than the existing ecolabels and comparable to the lowest (Dutch) directive. In the debate the user presented the view that although they go for the best environmental products, they need good performing products as well and therefore preferred to set "minimum requirements".

The conclusion from this example is that the message towards the public *(to go for perfect non-toxic and ready biodegradable lubricants)* is weakened due to the disagreements in the market and the compromise that had to be found. One might have the idea that it would have been better to present "the extremes" with only a limited amount of complying lubricants, than to present a compromise that is even questioned by critical co-workers within the suppliers' organisations themselves.

More strongly formulated, one might put forward the question "is it wise to form a technology transfer team consisting of partners with opposite interests (Dutch Polder model), or is it better to form a group that might present idealistic (extreme) ideas, that finds its opponents in practice"?

As pointed out before, this more uniform team would consist of partners who can form ad hoc coalitions that are not necessarily stable for the whole period. If it is concluded that the goals will not be met with a certain coalition, or that LLINCWA fails to form a stable coalition within a certain branch, the goals might be reached another way around with different coalitions.

The LLINCWA experience shows that, although the participation of suppliers led to a non-optimal bio lubricant definition, their participation gave an enormous impact to the developed LLINCWA activities. The participation of two important multinational lubricant suppliers in the LLINCWA team gave a large support to involve other suppliers at the national and international level and to commit them to a large extent to the LLINCWA goals. Their positive cooperation in national pilot projects gave the LLINCWA activities a great impulse. The not-establishment of a critical bio lubricants definition did not prevent the stimulation of ready biodegradable, non-toxic lubricants. In the contrary, for many applications "critical bio lubricants" were identified and it was shown that they could be applied without any harm.

### LLINCWA's coalitions for support

As has already been stated, a strong aspect in the LLINCWA approach is the ability of the team to form ad-hoc coalitions. Lubricants are used everywhere, in large and small equipment, professional and non-professional applications, indoor and outdoor, as loss lubrication or in contained situations. The amount of products is almost unlimited and is used by a heterogeneous type of users. In the already strongly limited focus of

LLINCWA on activities in the aquatic environment, user groups vary between the individually oriented inland skippers and highly organised water management organisations. Involved branch organisations vary from associations of inland skippers and inland skippers waste management organisations until the associations of recreational sailors and water skiers.

Additionally LLINCWA is dealing with the suppliers and their national and international organisations, different governmental organisations concerned with general product policy, but also with governmental organisations for (water) transport, for water quality, for working conditions, with semi-governmental organisations that deal with international agreements on the Rhine (like for example the International Rhine Commission) and last but not least with the European Parliament to find support for the LLINCWA goals.

Therefore LLINCWA did (or perhaps had to) chose for a combination of a structural and a flexible organisation of the promotional and technology transfer activities. Already in the design of the project the important role of national advice commissions (NACs) was foreseen. They have a strong institutional basis (governmental representatives but also representatives of associations of suppliers); they have regular meetings during the whole project period, giving advice on strategic items within the project. But, since the NACs are thought to advise on quite diverse topics during the whole span of the project it is difficult to motivate the individual members for the whole period. Therefore the mentioned interest groups were only represented to a limited extend in these NACs.

Some groups with whom LLINCWA co-operated, clearly favoured a market based stimulation trajectory. Others were more in favour of developments along the political, regulatory line. By being flexible in switching between alliances, LLINCWA was able to pursue both trajectories simultaneously.

Stimulated by the systematic approach suggested by the Accompanying Measure CLIP[33] LLINCWA developed thorough national marketing plans, identifying key actors in the different selected market sectors facilitating the formation of goal oriented ad hoc coalitions (see annex). A nice example in this respect is the ad hoc coalition in the Dutch recreational sector: together with the environmental organisation *Waterpakt*, two suppliers and a lubricant dealer LLINCWA was able to form a crystallisation point for a coalition of the ANWB (the Dutch Driving, Biking and Walking Tourist Bond), the Union of Water Boards, the Ministry of Transport, Public Works and Water Management, the NVVS (Dutch Association of Fishermen), the KNWV (Royal Dutch Water Sports Association), the NWB (Dutch Water Ski Association) and the branch organisation for water sports allied companies (HISWA). This group together stimulates the use of bio lubricants amongst recreational sailors and might be able realise a break through at the recreational side but also at the side of the suppliers and the trade to deliver bio lubricants in this market segment in better suitable packages.

---

[33]    Accompanying Measure CLIP within the Innovation Programme of the 5th Framework Programme.

The formation and stimulation of these ad hoc coalitions takes a lot of time and effort, but seen from the perspective of networking and diffusion of the message towards relevant user groups this flexible organisation gives the LLINCWA activities a large momentum.

**LLINCWA's project design**

The "simple" idea behind the design of the LLINCWA project workplan, is the following. Based on the assumption that good performing bio lubricants are available (or can be easily formulated) for the defined purposes, the *state-of-the-art* (best available techniques) can be defined. The state-of-the-art can be proven, by testing the products in relevant applications. This result, the proof, is disseminated with the aim that the product will be adopted due to the better environmental performance. Subsequently some form of institutionalisation might be realised by the establishment of *"a critical mass of users"*. If a stable critical mass could be established without legislation one might speak of a self sustaining innovation: the amount of (motivated) users is large enough to motivate other users as well and not less important, the suppliers will be stimulated to formulate and sell the requested products.

Nevertheless, within the available time of three years and solely based on convincing environmental and technical performance arguments LLINCWA was not able to create a critical mass of bio lubricant users in the application area aimed for: the inland and coastal waters. Although the time of three years is too short to organise and perform a well-balanced amount of thoroughly monitored pilot projects enough proof could be generated to show a perfect performance in many different applications. Promising successes are reached within the market sector of the water management, there is an unambiguous positive attitude towards the use of bio lubricants. Even the very difficult reachable inland skippers (they are always "on the road") show a positive attitude in individual talks. The change towards the actual purchasing of bio lubricants and a sustainable use takes more time, financial barriers are high and since lubricants are only a "by-product" in the production the bio lubricant has only a limited building capacity for the environmental image of the user. Their use in the machine or operation is essential, but during use you don't see them. As soon as you actually see them something is wrong (for example accidental spill). (At the same time however, it is a quite simple operation to show your environmental behaviour just by using bio lubricants.)

As a consequence a permanent sensitisation might be needed to motivate the user in the long term, for a sustainable responsible purchase policy. The risk for regressive behaviour remains, due to a well-known short environmental memory and critical economical behaviour. Additional instruments to stimulate or even force users to apply bio lubricants are essential. Other instruments are covenants, financial stimulation measures or environmental (product) regulation.

*These activities were foreseen in the workplan and focused on the development of governmental regulation and voluntarily initiatives like ecolabel. Again in this respect the three-years project time is extremely short to realise well-established finished end products. Awareness could be created amongst environmental national and European governmental authorities, a connection in running legislation or measures could be realised like for example the mentioning of biolubricants in the European "Directive*

*on laws, regulations and administrative provisions relating to the recreational sector",
that asks for the enforcement to use biolubricants in the (aquatic) recreational sector.
Direct success could be gained because connection could be realised to an already
running process. The same holds for example for the Dutch Vamil regulation, where
biolubricants could be accepted after ample negotiations. (Nevertheless, the elections
of a new (right-wing) parliament in 2002, simply resulted in an abolishment of the
Vamil regulation as a whole, with the thread that this will lead to a fall in use for
biolubricants. This type of political restoring measures turn out to be taken quite easily
not withstanding the negative environmental consequences.)*

Especially this last example shows the importance of reaching a critical mass in the
amount of users of biolubricants, a critical mass that is, relatively independent from
legal or financial stimulation, actively demanding for biolubricants (being in this way a
stimulating factor for lubricant suppliers to develop and supply biolubricants).

**Conclusion**

It can be concluded that the *Innovation programme*, that gives an opening for the
performance projects with a holistic approach, was an excellent programme for the
LLINCWA activities. Combining the different roles (the messenger (innovation
advocate), the supplier and the user) in the team does lead to a stronger market oriented
message. "Conflicting" items can be sorted out within team before the message is being
sent out, resulting in a committed and liable team able to defend the message in public.
At the same time, however, some conflicts cannot be prevented, leading to some extent
to a weakening of the intended environmental message. The conflict about the used
definition for biolubricants can only be solved when an agreement is reached with most
of the market players – although inevitably some losers will remain, for example those
who supply lubricants that fall outside the agreed biolubricants definition. A European
ecolabel could play a role in harmonising this definition.

The realisation of broad support for the biolubricants by establishing national
advise commissions, but also to go for ad hoc coalitions in specific cases, within a
project design that gives room for creative and "unforeseen" initiatives, proves to be
successful.

Although barriers are shown within the context of the non-food applications of
renewable raw materials especially concerning the generally significant higher
purchase price of these bio products and the fact that a reduction of price is not very
likely to be realised without explicit tax or other financial incentives, efforts must be
continued to realise a stable larger demand for biolubricants. Awareness and
environmental motivation is a key factor in this respect. Nevertheless trying to increase
the market demand only based on technical and environmental considerations is bound
to result in only a low scale use. A low scale biolubricants demand, insufficient for
suppliers to motivate the development of new specific products and a restriction to go
for a full market supply, will be the result. Therefore, to make biolubricants a real
success governmental (legal or tax) incentives are indispensable.

# Chapter 10

# Proceedings of the LLINCWA conference in Paris

The LLINCWA conference, which took place at TotalFinaElf headquarters in La Defense, Paris the 15th of November 2002, was organised by the LLINCWA consortium. The goal for organising this conference was to give a condensed overview of the successes reached so far by the LLINCWA project, to discuss some fundamental problems concerning the acceptance of biolubricants in practice and to take care for a further dissemination of the message: *"Biolubs are available; biolubs are liable; they even perform better than the "traditional" lubricants; there is an environmental need for substitution; substitution is technically and economically feasible"*:

## 10.1 INTRODUCTION OF BIOLUBRICANTS TO THE MARKET

The conference was opened by mr Hagenbach, the lubricant vice president marketing and technique lubricants of TotalFinaElf. His message concentrated on the strong need for the public to reduce its ecological footprint and a wise choice for biolubricants can give good contribution to this. LLINCWA focussed on the aquatic system, which is of course a key system in the environment, nevertheless, for the lubricants industry the use of biolubricants in transport, especially in automotive applications is even more interesting. With the use of biolubricants it will be possible to reduce the $NO_x$, CO and $SO_2$ emissions significantly due to improved lubrication properties. Life cycle studies can be of great help as an instrument to stir R&D.

Chairman of the day mr. Woydt from the German Bundesanstalt für Materialforshung und –prüfung confirmed the statement that the automotive industries is a key industry to stimulate the biolubricants market. The enormous use in this sector will be of a great relevance, creating a large sales and consequently a rationalising of purchase prices.

Mr Dohy from the French Ademe, did put emphasis on the environmental need to change over towards the use of biolubricants.

Mr. Van Broekhuizen presented some of the findings of the LLINCWA project. These can be found elsewhere in this LLINCWA report.

## 10.2    A HOLISTIC APPROACH TO INNOVATION

The European Commission was represented by mr. Haesen, presenting the view of the Commission on the process of innovation and the way innovation has to be understood as a holistic approach in the modern market[34]:

Innovation conjures up an image of a new technology passing from the realm of research into the commercial sphere, to provide benefits in the real world. But this simple linear model, with a technology provider in one Member State supplying end-users in others, does little to improve the wider capacity for innovation.

Efforts to support technology transfer have focused on technical issues; they have tended to deal with those organisations taking part directly in the technology transfer itself. In reality the innovation process affects a far wider range of organisations than just those directly involved. The widespread of Knowledge arising from the implementation of Innovation

Projects, launched under the European Commission's Fifth Framework Programme, contribute to the development in Europe of an "innovation culture" and demonstrates the applicability of innovation in particular sectors of activity. Over the last years support has increasingly been targeted at complex innovation structures, aiming to establish a culture within partner organisations of identifying and addressing analogous obstacles. A number of non-technical barriers (inadequate management capacity, bad communication, and poorly understood end-user requirements) can hamper effective technology transfer and adoption.

Links forged between the core knowledge providers and the beneficiaries by "gearing" intermediaries such as local authorities or umbrella organisations have proven to be successful. Involving a wider group of stakeholders, especially if this occurs at the project conception stage produces a better understanding of market barriers and market potential. Projects, which have systematically addressed these pitfalls in parallel with more conventional technical work, have derived substantial practical benefits in terms of speed, range and sustainability of the resulting innovation.

The analysis of good practice indicated that the successful implementation of technology transfer is very much depending on the correct aggregation of skills, including technology, market and management skills in exploring relationships within the consortium and outside the boundaries of the project. If most projects are initially built on the individual interest of participating organisations, the successful innovation is based on sharing the knowledge available in the consortium and the added value of community co-operation.

---

[34]   **A holistic approach to Innovation,** Guido Haesen, Francisco Fernandez Fernandez, Jean-Claude Venchiarutti.
European Commission, DG Enterprise, Directorate Innovation & SMEs *This contribution to the LLINCWA International Conference is based on a first set of lessons learned through the implementation of Innovation Projects and related Accompanying Measures of the Fifth Framework Programme of the EC.*

## 10.3    OEMS AND THE USE OF BIOLUBRICANTS

Mr. Kirsch from Bosch Rexroth AG in Germany presented the central role of original equipment manufacturers in directing the user to use biolubricants: Environmentally-safe fluids for hydraulics used in civil engineering, the long way to go, and the demands an OEM is able to formulate.

The majority of hydraulic units used in civil engineering are operated with pressure fluids based on mineral oil. Most civil engineering projects are installed near or immediately next to bodies of water, therefore, any leakage signifies danger for the environment.

We avoid this danger with increasingly safe hydraulic drives. However, growing environmental awareness and stricter laws are demanding more and more environmentally safe hydraulic fluids.

Today, the manufacturers of fluids and hydraulic drives have to accept this challenge.

The following application areas specifically require the use of environmentally safe fluids:
- All kinds of civil engineering applications as ship locks, dams, movable bridges etc.
- Offshore and marine hydraulic applications
- Hydraulics for all kinds of food preparing and packing
- Earth moving machines

That's to say in all areas where leaking oil can penetrate the water or the ground or where it can get in contact with food, environmentally safe fluids are to be used. Unfortunately, neither there have been laws nor standards which clearly define the attribute „environmentally safe" hydraulic fluid, and there are no directions about what to do in case of an accident, such as leakage of large quantities of these fluids. If there is an accident, the same guidelines apply as with mineral oil because even environmentally safe fluids have to be handled carefully and they cannot be left behind in the environment after they have leaked. However, ground cleanup due to pollution will be much more cheaper than with mineral oil.

How does the application of environmentally-safe fluids affect the design of hydraulic drives?

*Use of native-based oils:*

As mentioned before, native vegetable-based oils are not very suitable for civil engineering projects at the present time, because of their adverse reaction to low temperatures. In only one of our projects we have used rapeseed oil as a hydraulic medium. It was done on a bridge application in Florida and we had negative results.

First we found gumming from the fluid of the piston rod and, because of bad maintenance of the fluid, the hydraulic drive suffered a number of failures. In the end, after approximately 3 years, we had to change from the rapeseed oil to a petroleum-based one.

*Use of polyglycols:*
Due to the mentioned adverse properties of polyglycols the following measures in the design of hydraulic drives have to be taken:
- Compatible paints or stainless steel reservoirs have to be used. Reservoirs made from aluminum are possible as well.
- Assembly / installation of the power units and cylinders without any grease or petroleum-based oil. The power units and cylinders have to be tested at the plant with polyglycol.
- Do not use any seals made of polyurethane
- Check all plastic materials for compatibility
- Prevent water from leaking into the reservoir by means of an air humidifier

*Application of synthetic esters*
The characteristics affecting the design of hydraulic drives are more or less the same as with polyglycols and therefore the following measures in the design of hydraulic drives have to be taken:
- Compatible paints have to be used or for instance stainless steel reservoirs. Also possible are reservoirs made from aluminum.
- The power units and the cylinders have to be tested at the plant with synthetic esters in order to flush the residual grease and petroleum-based oil from previous assembling
- Do not use any seals made of polyurethane
- Check all plastic materials and metals for compatibility
- Prevent water from leaking into the reservoir by means of air humidifiers

Bosch Rexroth began at a very early stage to test environmentally safe fluids for their possible use with hydraulic controls. In the 1970's we tested valves, pumps and other hydraulic components on the test stand using various environmentally safe fluids. In 1980 we had completed our tests far enough in order to be able to equip the first small vessel lock on the river Main in Germany with an environmentally safe fluid. The fluid we selected was a polyglycol made by the Company FRAGOL. At that time, we were already aware of the fact that glycols corrode standard paints and since we weren't aware of any paints compatible with these fluids, we hot-galvanized the tank externally and internally. We assembled/installed the power unit and the cylinders in a normal way, e.g. as if we were using petroleum-based oil. We constantly monitored the hydraulic system for this small vessel lock and tested it at certain intervals. After one year, after the system went through one summer and one winter, we came to the following conclusions:

We found a jelly-like substance in the filter and on the surface of the fluid. After a chemical analysis of this substance, we found that it consisted of petroleum-based grease and oil, and that it originated from the use of grease and oils during assembly/installation of the power units and cylinders. During analysis of the fluid we found considerable amounts of zinc, which were there due to a corrosion of the tank coating. Despite the jelly-like substane in the filter and on the surface of the fluid, as well as the high amount of zinc in the fluid, the system ran an entire year without any interference.

After this first year, we completely replaced the fluid and flushed the system in order to flush out the residue of petroleum-based oils and grease. Now the system has been running for over 22 years without any problems and there were no further jelly-like substances to be found.

What we have not been able to prevent is the washing out of the zinc coating in the tank. However, since the process was one of hot galvanizing with a coating thickness greater than 200 micron, we established theoretically that the corrosion protection through an increasingly thinner zinc coat will last another 20 years.

Based on this knowledge, we equipped additional civil engineering projects, e.g. locks and dams, with this fluid. However, we used stainless steel reservoirs. In cooperation with various paint manufacturers, we moreover developed paint systems that are resistant to polyglycols. We have been using reservoirs with these paints since 1986.

In general, we can say that approximately 200 units designed by us are in operation with polyglycols.

Because of the increase in demand for these polyglycols, the various petroleum oil companies have made quite a lot of efforts to develop new, environmentally-safe fluids. They developed Synthetic Esters which:

a)    can be used in hydraulic systems and which are

b)    identified as environmentally-safe fluids

We have tested our components on the test stand with these synthetic esters, and since March 1992 the first lock has been in operation on the river Danube with synthetic esters.

To-date, the system has been running without any problems because synthetic esters have shown properties similar to those of polyglycols, and we could use our experiences with polyglycols when we designed these systems. Other projects like power plant Greifenstein on Danube river were following.

All the above-mentioned experience with polyglycols have been gained in new applications. We have only changed the medium at already built installations at the Heidelberg Sluice, together with the Water and Ship Authority of Heidelberg. The main reason for this was that the applied coating in the existing units weren't resistant against these environmentally friendly mediums, and replacement entails high costs.

At the Heidelberg lock, synthetic ester from SHELL was applied. Since we were not familiar with the coating of the oil reservoir, new oil reservoirs from stainless steel were manufactured. The hydraulic components were partially renewed or completely cleaned of petroleum-based oil. The unit is operating problem-free since the summer of 1992. After a period of a one year (summer and winter operation) we tested the unit. No negative influences regarding our components or the function of the complete unit were found. However, we are sure that in the future and in the course of further growing environmental consciousness, more and more units will be converted.

We have equipped the Detzem lock on the Mosel river with a synthetic ester of Fragol since May, 1995. This is a new installation. The power units for the modernization are equipped with gear pumps. We were sceptical, that the fluid of the gear pumps could destroy the plain bearings. However, the tests carried out on our test stand in cooperation with the bearing and fluid manufacturers did not confirm this. The installation runs trouble-free to the present day.

After this we equipped a further 10 units using this medium, among others the recently openend Ems Storm Flood Barrier, which all run trouble-free.

Based on the more than 20 years of experience with environmentally-safe fluids, it can be said that today, manufacturers of hydraulic equipment should have no problems operating systems with the presently available environmentally safe fluids.

There are no additional costs for new systems, except the higher price for the fluid. It is more costly to switch current systems over to these fluids, mainly because of the existing paint, but it can be done. We are confident that the development processes of the environmentally safe fluids have not yet been completed. We have established a catalogue with all the requirements for these fluids and we forwarded it to the interested fluid manufacturers.

The major requirements for new fluids are:
1. Good lubrication
2. No corrosion of materials, i.e. paint coats
3. Good viscosity-temperature characteristics
4. High thermal and oxidation features
5. Low compressibility
6. Low foaming tendency
7. High density
8. Good thermal dissipation
9. Non-poisonous
10. Low costs
11. Easy availability
12. Low maintenance
13. Disposal without problems

The new and improved fluids will surely change the design requirements for hydraulic systems, which will have to be addressed by the manufacturers of hydraulic controls. Improved hydraulic fluids, regardless of how environmentally-safe they are, do not diminish the responsibility of the project engineers and suppliers of hydraulic systems to design hydraulic drives which are safe and leakage free acc. to DIN 19704, since one thing should be kept in mind by all those who are dealing with hydraulic fluids: Absolutely environmentally safe fluid can only be clean water!

## 10.4   BIOLUBRICANTS AND THE USER

The point of view of making the choice to substitute conventional lubricants by biolubricants, from the point of view of the user was presented by mr. Maldet from TIWAG: Experiences with environmentally acceptable lubricants and hydraulic fluids in TIWAG -Tyrolese hydroelectric power stations. An extensive overview of their experiences is presented in chapter 8.

## 10.5    BIOLUBICANTS AND THE SUPPLIER

Mr Bush from Fuchs Lubritech GmbH presented the position of a pro-active supplier strongly motivated to develop and market biolubricants. He formulated the main questions a suppliers formulates for himself to decide
- Do our customers need/want/ask for biolubricants?
- • What is the market potential? What is achievable?
- • Shall we invest in R&D, production and marketing of biolubricants?
- • Time to market?
- • What is the return of invest?

After a "simple" yes to the first question, they base their judgement on the market potential on a study from Frost and Sullivan :  In  1999 a potential is estimated of 750,000 - 1,000,000 t/a with a market value of 231 million € with a prognosis of increasing  664 million € in 2006. This will be influenced by stricter environmental regulations,  technical requirements (for example emission limits), promoting programs (Germany). In the long term it is expected that more than 90% of all lubricants might be ready biodegradable.

Investment in R&D and production and marketing of biolubricants is strongly determined by the actual market and of course the environmental demand. For Germany the loss of lubricants in the environment from different lubricant-sources is represented in the following figure. Directly spilled in the environment are total loss lubricants. Their relative market share is represented in the following figure.

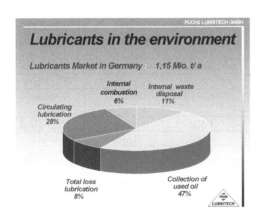 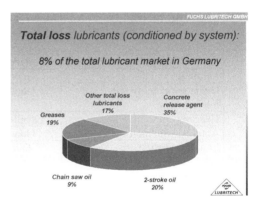

Fig. 10.1

The time from development of the biolubricant towards acceptance and final use is experienced as too long, much too long. To shorten the 'time-to-market' process several initiatives can be taken: showing customer benefits, add value to our customers business, inform objectively about the pros' & cons' of biolubricants and to activate public awareness and support. Important is the initiative from the government in this respect. The German initiative from the ministry of agriculture (extensively described in chapter 5) is a good example. Additionally ecolabels can do good work, but need strongly a harmonised approach. At this moment there are strongly differing criteria to denominate biolubricants. Therefore the development of a European ecolabel is strongly favoured.

Mr Westelynck from TotalFinaElf strongly focussed his presentation on the need to select the proper additives for the biolubricants and the commonly used base fluids in biolubricants.

Table 10.1

| Lubricants | Hydraulic | Gear | Compressor | Turbine | Grease |
|---|---|---|---|---|---|
| Additives: | | | | | |
| Anti-foam | x | x | x | x | |
| Anti -rust | x | x | x | x | x |
| Anti-corrosion | x | x | x | x | x |
| Anti-oxidant | x | x | x | x | x |
| Anti-wear | x | x | x | x | x |
| Extreme-pres | x | x | | | |
| Viscosity modifier | (x) | | | | |

Although the performance of lubricants, is dependend on the base fluid as well as the additives that are used in general a comparison can be made of the performances of the different types of lubricants:

Table 10.2

| PERFORMANCE | Mineral oil | Vegetable oil | Synthetic ester | Polyglycol |
|---|---|---|---|---|
| Viscosity range | Wide | Limited | Low only | Very wide |
| Lubricity | Good | Excellent | Excellent | Excellent |
| Oxidation stability with inhibitor | Good | Poor - Fair | Good – Excellent | Good |
| Hydrolytic stability | Excellent | Poor | Fair | Very good |
| Anti rust with inhibitor | Good | Good | Good | Good |
| Low temperature use | Fair | Poor | Good - Excellent | Excellent |
| High temperature use | Fair | Poor | Good | Good |
| Low volatility | Fair | Excellent | Excellent | Good |
| Compatibility (Seals, paints, varnishes…) | Excellent | Good | Fair | Poor |

## 10.6   WORKSHOPS

Four workshops were organised:
- Inland marine & recreational shipping
- Water management and hydroelectric power plants
- Mobile equipment
- Harmonising the biolubricant definition

The conclusions from the workshops can be summarised as follows:

Biodegradable non-toxic lubricants (biolubricants) *must* be used in the majority of applications on and around the water: on ships, in locks and sluices, in mobile equipment, in hydroelectric power plants. The aquatic pollution caused by the use of mineral oil based lubricants is unacceptable since in most of the application a suitable ready biodegradable, non-toxic alternative is available. The use of biolubricants is technically and operationally feasible, economically viable (and often profitable) and most of all: environmentally benign and necessary. Although plenty of users have positive experience with biolubricants, the majority of the users is still unaware, or is deterred by higher prices, by bad experiences with older, immature products or by the lacking guarantee of original equipment manufacturers (OEM's) in case users switch to biolubricants.

*More efforts needed, particularly by authorities*

Much more effort must be dedicated, therefore, to inform users of the availability, quality and advantages of biolubricants. Co-operation can and should be established between users, OEM's and suppliers to ensure the warranted use of biolubricants in all applications – a co-operation that all parties that were present said they are willing to engage in. Moreover, a distinct need was identified for more government initiatives, preferably at EU level, which can either be of a financial nature (like the subsidies at present in DE; the tax advantages in NL, or taxes / penalties), of a regulatory nature (like on the German Lakes or in Scandinavia) or of an exemplary nature (like the systematic purchasing of biolubricants by German public bodies). Such initiatives are deemed indispensable for a serious growth of biolub market share to occur.

*Towards common biolub criteria*

The first requirement for such government initiatives is the existence of a harmonised definition of biolubricants – whereas at present all eco-labels, national policies and OEM requirements refer to different criteria for biolubricants and different testing methods. All experts agree that such a harmonised definition is urgently needed and feasible where toxicity and biodegradability are concerned. Common criteria for content of renewable raw materials require more in-depth exploration.

Most of the participants were in line with the statement laid down in the background document and believed that biolubricants can achieve the same - and even higher - performance as standard products, if they are fulfilling certain technical standards, e.g. those defined in ISO 15380. Therefore, one important element of a biolubricant should be that it fulfills these technical minimum requirements. To meet these technical requirements additives are inevitable, but should be limited to a total a low content.

Most of the participants agreed as well that the establishment of a European Eco-label would be desirable in order to improve visibility and prevent confusion among consumers. However, there was less unanimity about the exact requirements. The discussion concentrated mainly on the question whether certain minimum requirements with regard to the renewability of the lubricant should be included in the eco-labeling criteria. Opponents of this view argued that the inclusion of a 'renewability' requirement would lead to the exclusion of products which are as biodegradable and as non-toxic as the renewable ones. A majority could however agree on the notion that lubricants formulated from at least 50% renewable materials are well available for a wide range of applications.

Another point of discussion was the biodegradability and toxicity of biolubricants. One important requirement for biolubricants should be that they are easily biodegradable. In this regard it is reasonable to distinguish between the base oils and the additives. The base oils should each be easily biodegradable. This is to be sure that highly biodegradable ones do not mask non-degradable components. It would be however unrealistic to require that the additives present in the lubricant should also be easily biodegradable. The second important aspect of a biolubricant is that a biolubricant should not be dangerous (= toxic) to the environment. A pragmatic proposal put forward during the workshop was that a biolubricant should not be classified according to the Dangerous Preparation Directive, 99/45/EEC.

Finally, some participants expressed the view that the establishment of an ecolabel would not be sufficient in order to increase the market share of biolubricants if not accompanied by some kind of financial incentives.

# PART 5

# FUTURE FOR LUBRICANTS

# Chapter 11

# Life after LLINCWA

The attention of most of the LLINCWA-groups to continue biolubricants-oriented activities does not stop with the ending of the innovation project. Most of the groups will continue to stimulate the use of biolubricants in one or the other way. A summary of different initiatives is given below.

### INPT-biolubricants activities

The French LLINCWA partner INPT, located in Toulouse, takes part in projects involving biolubricants and more generally eco-compatible formulations. These programmes supported by AGRICE/ADEME deal with different activities fields such as agriculture, forestry or paintings.

INPT is also involved in the conception of a standard test for the assessment of the lubricant biodegradability in soil (partnership with: AFNOR, ONIDOL, ADEME)

### ValBiom -biolubricants activities

The Belgian LLINCWA partner ValBiom, located in Gembloux, will continue the monitoring of the LLINCWA demonstrations and pilots. Some are not yet finished and as agreed in the signed agreement with both the user and the supplier, they continue their monitoring activities in analysing samples and discussing the final results of biolubricants use at the Ronquière-rollers and the Ath-locks. Other pilots are still running, but these are only neither loss lubrication applications that do not need a close followed up nor analysis.

Beside the water activities, other biolubricants demonstration projects are running in tractors and mobile machines since 1997.

Successful was the setting up of the Belgian national advice group. This group will continue its activities coordinated by ValBiom as working group on "plant-oil based lubricants ". Their goal is to keep the Belgian stakeholders informed with new developments on biolubs.

### Tekniker -biolubricants activities

Tekniker, the Spanish LLINCWA partner has already a long history on R&D in the field of biolubricants. Besides their "normal" day-to-day research activities in this field, they took the initiative to develop a proposal for a European Project for the development of compatible environmentally friendly oils trying to substitute harmful additives by the effect of triboreactive materials in engines and automotive gear oils. Based on the conclusions of the LLINCWA-project and together with other LLINCWA partners a large integrated project to be applied with the 6th Framework programme is under development.

### ISSUS –biolubricant activities

The German LLINCWA partner ISSUS, located in Hamburg, has a somewhat exceptional role within the LLINCWA team. Responsible for the engineering education activities of the Hamburg University of Applied Sciences they have a direct role in the education of students, in later life active as marine engineer. And since the German bio-lubricant-market is the biggest in the EU, and since biolubricants still suffer a bad image in a wide range of public consciousness ISSUS took the initiative to proof the LLINCWA results in two ways:

- To expand the amount of information about bio-lubricants in the engineering education activities of the Hamburg University of Applied Sciences. This did not only relate to the POL (patrol, oil and lubrication) lectures for marine engineers at ISSUS, but the German advices group has also shared the LLINCWA- findings with the department of mechanical engineering at Hamburg's University of Applied Sciences and of the Technical University of Hamburg-Harburg. This is a foundation-stone for a better acceptance of bio-lubricants in the future.
- The German LLINCWA website will get a longer life, ISSUS intends to provide and to manage the site in the future: <http://issus.susan.fh-hamburg.de/iss_ web/projekte/llincwa/index.html>

### Fuchs – biolucant activities

The German LLINCWA partner Fuchs Lubritech has its core business in lubricants and will continue its activities in R&D and Promotion of Biolubricants.

For many years Fuchs Lubritech has been investing in research and development of environmentally harmless lubricants and their promotion on a global basis. The products resulting from this development are known as Fuchs Lubritech ECO-Line products. In the product literature reference is made to the LLINCWA project which may help to promote both the ideas and targets of the project on one side as well as Fuchs Lubritech's company policy of being conscious of environmental and economical matters. Spurred by the LLINCWA project, the research and development activities in the field of environmentally harmless lubricants have even been enhanced so that by today, a complete range of these products can be offered. Some typical examples are:

Environmentally harmless EP multi - purpose greases like *STABYL ECO S 12 G*, which contains solid lubricants. It is manufactured with rapidly biodegradable raw materials. *Stabyl ECO S 12 G* is applied to all plain roller bearings, even when highly stressed, for which multi-purpose greases are specified, especially if mineral oil based products could pollute the ground or water. Because of its excellent rheological behaviour at low temperatures, the grease can also be used at low ambient temperatures.

Environmentally harmless synthetic ester-based gear oils like the *GEARMASTER ECO –oils*. These oils offer outstanding protection against fretting and micropitting. They reduce friction and wear and optimize the start behaviour of gearboxes, especially at low temperatures. *GEARMASTER ECO* -oils are recommended for the lubrication of spur, bevel, planetary and worm gears, particularly in the wind-power-industry, as well as in food processing and packing industries, but also for machinery operated in protected water zones.

Environmentally harmless hydraulic fluids like *ECO-HYD S PLUS*. This product is based on a synthetic ester, largely applicable and able to replace mineral oil based types ranging in the viscosity classifications ISO VG 22 to ISO VG 68. *ECO-HYD S PLUS* shows an excellent low-temperature behaviour as well as an outstanding temperature and ageing stability. The reserve performance of *ECO-HYD S PLUS* surmounts those of conventional mineral oil based hydraulic oils.

Moreover, Fuchs Lubritech  is developing lubricants for railroad applications according to the very strict Scandinavian standards.

The Lubritech ECO-Line brochure gives a survey of the environmentally harmless products available in Fuchs Lubritech's product portfolio.

### TotalFinaElf - biolubricants activities

The French LLINCWA partner TotalFinaElf, located in Paris, has its core business in lubricants as well. They will continue pilot tests that were launched during LLINCWA will be monitored carefully trying to formulate an answer to some specific questions, to be sure that the started projects will end up as a durable success.
- Several tests being run in the Netherlands & Belgium on fishing and inland shipping boats. Biolubricant sales have been already achieved in this activity field, mainly concerning stern tubes and hydraulic systems.
- Tests made in relation with the French VNF organisation, both a user and a prescription authority for the use of biolubricants. TotalFinaElf will pursue their effort to make VNF to prefer the biolubricant solutions and to expand their use in other equipment. The same approach will be followed by the tests made with grease in hydroelectric power stations.

TotalFinaElf's involvement in the biolubricant development is not only focused on the Llincwa project and they have several other developments:
- Application for farming equipment: Under development is a multifunctional biodegradable fluid that will become available for farming equipment.
- Biodegradable and food grade product: This new concept is presently tested in crop machines and will soon be available on the market
- Bio-hydraulic fluid is also presently tested on Vine tractors
- Biodegradable chainsaw oil: Composition and performance level of our products are actually studied to make them even better.

In general the TotaFinaElf involvement in LLINCWA and in other biolubricant development projects have lead to the development of a specific brochure that will be used in the TFE-commercial organisation to support the sales of biodegradable lubricants; an English version of this document is appended.

### Hoogheemraadschap van Rijnland – biolubricant activities

The Dutch Waterboard Hoogheemraadschap van Rijnland, located in Leiden, was one of the first larger-scale users of biolubricants amongst Dutch water boards, already before LLINCWA started her activities. Their participation in LLINCWA led to a further acceptance of these products within the company. Specific benefits of biolubricants in the large water management systems will be further studied, to even get a stronger proof for the environmental benefits. This concerns tests to measure the noise reduction in running machines. Due to a better performance of biolubricants first tests show a noise reduction, which will be an excellent answer to complaint of neighbours living nearby pumping stations. Comparable tests are carried out to measure vibration reduction and the reduction of energy consumption due to the use of biolubricants.

New goals formulated for Hoogheemraadschap Rijnland:
- Adapt all internal lubrication manuals to the use of biolubricants
- Changing of used lubricants only after monitoring samples in stead of after a fixed time period
- Research on energy reduction, reduction operational temperature and noise reduction due to use of biolubricants, determining lubricant film thickness related to load and play
- Environmental benchmarking of biolubricants for waterboards
- Inventory of used lubricants for all waterboards: total amount purchased/ percentage biolubricants/ use / waste / costs
- FZG

### IVAM – Chemical Risks - biolubricant activities

The Dutch research and consultancy group IVAM, department of Chemical Risks continues its lubricant activities on several levels.
- During LLINCWA a close cooperation has been developed with ELGI, the European Lubricating Grease Institute. With them IVAM wrote the book "Health and environmental hazards of commonly used additives in lubricants". This book, that presents environmental and toxicity data available in the open literature, was subject to an intense discussion with additive suppliers. This discussion on additive-data will be continued.
- Good contacts have been established with the national platform on diffuse (aquatic) pollution. Due to the LLINCWA activities lubricants are identified as one of the sources of diffuse aquatic pollution. In this platform participate the environmental movement and many provinces. Within this framework plans are under development to organise special project ship on the introduction of biolubricants in water-management activities. This plan will be further developed and specific projects will be set up to further increase the awareness on the existence of biolubricants and to increase the use. Especially with the Waterboard Amstel, Gooi

en Vechtstreek agreements for an awareness campaign to be organised the end of 2003, beginning 2004 are settled.

- Within the framework of the Dutch White Paper (environmental chemical substances policy) a national project is being carried out to identify the lack of data on chemicals within the lubricant branch. Within this project, chaired by Shell, twelve organisations do participate: raw material producers (Uniqema, King Industries, Coolant Contol), lubricant manufacturers (Shell, Castrol, Axel Christiernsson, Fuchs Laagland), an OEM (Spaans Babcock), lubricant users (Waterboard DWR, shipping organisation BBZ) as well as the trade union (FNV Bondgenoten) and the environmental movement (Stichting Natuur en Milieu). IVAM coordinates this project and carries out the work.
- With the Dutch OEM IHC manufacturing of screw axe bearings a cooperation is being set up to test and design biolubricants for their systems.
- With the former LLINCWA partners, under the overall coordinaton of Tekniker, IVAM coordinates one one of the four workpackages of the integrated project EuroEcolubs, that is being applied within the first Call for proposals in the 6th EU-Framework Programme.

# Chapter 12

# Conclusions, limitations, opportunities and challenges

After three years of LLINCWA activities to stimulate the awareness on the existence of biolubricants, testing and demonstration of the performance of these products it can be concluded that biolubricants exist and that they form an excellent answer to the environmental demand for a reduction of the diffuse aquatic pollution with mineral oil. They exist but they are not available everywhere.

## *They are available, but....*

- They are available...... on specific request, but not at all suppliers for all purposes. Many applications demand tailor-made lubricants, but fact is that most suppliers do not sell tailor-made products for all the requested applications. Their specialism is largely based on their (commercial) market analyses, with the result that in practice certain applications will have to comply with "over-qualified" developed products, if the user insists buying the lubricant with its usual supplier. Which might mean a proportional purchase price. Or it means that the user has to find another supplier that sells the requested product.

- They are available...... but it is not always well-recognisable which performance is needed for the specific application. Frequently an over-qualified product is advised while a just-fit-for-use product can do the job as well, and seen from a commercial point of view even better. Just-fit-for-use means almost always a lower purchase price than the high-demand products.
  But recognisability means as well the ability to distinguish *bio*lubricants from products with a lower environmental performance. Several national ecolabels (or ecostandards) have tried to clarify some of this confusion, but since the formulated criteria are not identical, confusion remains, and only a minority of the available products actually are labelled with a (national) ecolabel. By developing a classification system LLINCWA tried to solve some of this confusion, but the strong need for a EU-harmonised definition for biolubricants remains. The operationally defined *set of minimum* requirements LLINCWA used to distinguish biolubricants is confusing as well. The message towards the public *to go for perfect non-toxic and ready biodegradable lubricants* is weakened due to the disagreements in the market and the compromise that had to be found about the used definition. One might have the idea that it would have been better to present "the extremes" with only a limited amount of complying lubricants, than to present a compromise that is even questioned by critical co-workers within the suppliers organisations themselves.
  Lubricant manufacturers strongly need the tool (a standard) for their product development. A European ecolabel that hopefully will be developed after LLINCWA might solve this definition question.

- They are available…… but not always in the requested package size. Biolubricants for the recreational boating market do not differ fundamentally from the professional market products, but they are not available in small quantity packages (1-2 kg) in the shops in the yacht-basin. The user market seems to be too small to be commercially attractive for packaging departments of suppliers. (And on the other hand, as an anecdotic example, the introduction of smaller barrels, a reduced size from 40 kg barrels to smaller 20 kg barrels, in order to comply with occupational health demands, failed because the professional users preferred handling the lubricant-vessels with a forklift).

- They are available…… in environmentally active countries, but in countries with a low pressure on a reduction of the environmental pollution they are absent. For the Spanish LLINCWA pilot projects biolubricants had to be imported separately. Supplier's turn out not be pro-active in environmental respect.
  The lubricants market is a strongly demand driven market but if a user doesn't know that they exist he won't ask for them. One might expect a rich market for niche players, but up till now only a minority of niche players amongst suppliers do play a role in taking the lead in developing an active marketing strategy for this environmentally beneficial product group. As a consequence unawareness amongst users is the result.

- They are available…… for may applications, but due to existing guarantee clauses on the used equipment and despite the low substitution-risk the user is not really stimulated to experiment with biolubricants. The guarantee might get lost if the machine would be damaged after substituting with biolubricants. The OEM (original equipment manufacturer) plays a crucial role in this respect, and should be strongly stimulated to test biolubricants in his equipment, or in setting up R&D projects to design biolubricants compatible equipment (complementary to the lubricants suppliers' R&D of developing equipment compatible biolubricants).

There is only a minority of applications where biolubricants cannot be applied and their performance in general is as good as "traditional" lubricants. Due to their physical properties their lubricating performance sometimes is even better than mineral oil products. With the development of synthetic esters (in general largely derived from renewable vegetable oils) biolubricants could even be further improved. Modern development shows the evolution of base fluids with a lower need for specific additives, resulting in a further reduction of the (potential) environmental pollution.

### *Loss and lost lubrication and their environmental burden*

LLINCWA focussed the activities on the use of biolubricants applied in and around the inland and coastal waters. Inland applications got the main attention. To structure (but also to limit) the activities four main types of lubricant applications were distinguished: inland shipping, water management applications, hydroelectric power generation and recreational sailing activities. In these applications hydraulic fluids, gear oils, greases and two-stroke oils were taken into account. Furthermore, two application types were distinguished: loss lubrication and lost lubrication. Loss lubrication is found in applications that use open systems, in direct contact with the environment, mostly using lubricants especially designed to be spilled in the environment during use (like for example screw axe lubrication, many greases and two-stroke oils). With lost lubrication accidental spills from closed systems are meant; spills that occur due to for example breakage of hoses in hydraulic systems.

*Significant use*
Educated guesses estimate a total yearly use of 5 million tons of lubricants in all applications: land and water in Europe. Suppliers do not distinguish between land and aquatic use, which makes it impossible to make a clear estimation of the yearly lubricants use in the aquatic area. Around 45% of these lubricants is being spilled somewhere in the environment: deliberately or accidentally, showing the enormous need to stimulate the use of biodegradable, non-toxic lubricants.

In the water-connected areas only an extreme minority of the actually used lubricants can be identified as *bio*lubricants, a use of generally lower that 1-5% even after the 3 years of LLINCWA activities. As a consequence inland and coastal water activities remain responsible for an enormous aquatic pollution with mineral oils and their additives.

*Significant emissions*
Measurements show a total load on mineral oil of 1032 tons/year (1997) entering the Netherlands as pollution in Rhine water, on a total pollutant load of more than 5.000 tons/year. This means a load of more than 20% due to mineral oil originating from all sources fuels, machine oils, spills from metal working fluids, lubricants etc. Added to this are emissions in the Netherlands itself. A calculation of the actual spill caused by inland shipping in the Netherlands shows a significant contribution to the total mineral oil load. (Loss) lubrication of the screw axe and rudder system of the Dutch inland fleet does contribute annually around 300 ton to the actual aquatic pollution with mineral oils. In the past the total oil spill of the inland fleet was considerably larger. With the separate collection of bilge water a strong reduction could be realised. The recreational sailing activities in the Netherlands, with their screw axes and still a high amount of 2-stroke outboard engines, yearly contribute a significant amount of 83 tons mineral oils, mainly emitted in the more environmentally sensitive areas.

Additional to these sources are the mentioned structural and accidental sources from inland water management, hydroelectric power generation and of course many on-land activities with a discharge to the fresh waters. Unfortunately it was not possible to estimate their separate to the aquatic pollution.

*Biodegradability and toxicity*
As a general statement one can say that the toxicity of (bio)lubricants is determined by the additives used, while the biodegradability is determined by the base fluids.

The main environmental problems with mineral oil used in lubricants are highlighted in its physical effect of staining essential organs and its low biodegradation (both aerobic and anaerobic). According to the EU criteria its acute aquatic toxicity is too low to classify many mineral oil distillates as hazardous for the (aquatic) environment. In contrast to mineral oil lubricants biolubricants, like vegetable oils, synthetic esters and some polyglycols show ready biodegradability under aerobic as well as anaerobic conditions. The toxicity of the oils and esters is low. Due to their ready biodegradability staining effects due to biolubricants are not likely to occur and long-term effects can be ruled out.

### Experiences with perfect performing biolubricants

LLINCWA did set up pilot projects where traditional mineral oil based lubricants were substituted by biolubricants. In all mentioned areas interested companies or individual users (especially inland ships) could be found to experience biolubricants as an environmentally beneficial alternative. Substitution is an operation that needs a careful assessment of the equipment, compatibility of the biolubricant with the materials in the equipment and the operation conditions. A close cooperation with the lubricant supplier, and in some cases with the OEM (original equipment manufacturer) is indispensable for selecting and testing the proper lubricant. In almost all pilot projects mineral oil products could successfully be replaced by biolubricants. A skilled pre-selection of the right biolubricant is essential. For all selected applications a biolubricant could be found with good performance. Laboratory tests of applied biolubricants show a good performance, and although used biolubricants show an increased toxicity this does not exceed environmental emission standards. A careful assessment of observed failures proves to be essential since initial associations with a lower performance (for example an increased environmental spill in a screw axe system after substitution) may turn out to be caused by existing (previous) equipment deficiencies.

### Barriers for successful market introduction

Nevertheless, despite all the positive signs of good technical performance, beneficial environmental behaviour, lower human toxicity and a relatively trouble-free substitution, the substitution towards biolubricants in practice finds many barriers in its way.

The predominant barrier is unmistakable the purchase price of biolubricants. They are significantly more expensive than equally performing mineral oil based lubricants. Sometimes only slightly for low-demanding applications, but for high demand purposes biolubricants may even be two to five times as expensive. Nevertheless, total cost price calculations may show an equal or sometimes even lower price for lifetime biolubricants used in closed systems, due to a longer lasting usability. Unfortunately these long-life arguments do not apply to loss lubricants and the price difference is not by a longer depreciation time. And long-term visions are surely not rooted in all companies: short-term visions dominate. It frequently happens that strongly environmentally motivated engineers, in their environmentally responsible choice for biolubricants, are overruled by their purchasing department, simply to go for cheaper products.

The high price even combined with an environmental awareness is an almost unfeasible step to take and additional stimulating measures have to be taken.

### *Overcoming the barriers*

It is concluded that voluntary measures alone will not do. Governmental measures that strictly aim at voluntary substitution will not provide the necessary impetus. Labels and standards are already in place (although a common European ecolabel is still missing) and have not been effective in themselves to reach a voluntary critical mass of users. (Nevertheless a positive aspect of these labels and standards is the stimulus it gives for lubricant manufacturers for product R&D).

Also, management systems and certification schemes will miss the target they aim for to a large extent, given the dominance of small and medium-sized enterprises, self-employed and public lubricant users on and around inland and coastal waters.

Measures aiming directly at a price reduction of biolubs (relative to mineral oil based lubricants) are highly favourable. Tax measures can be thought of here, but also the German FNR initiative provides an interesting example of modes of financial stimulation. Indirect measures via financial stimulation of investments in machines using biolubs (like the Dutch VAMIL approach) can be fruitful in some cases. The other high potential type of government intervention is regulated use. Examples from the Swiss and German lakes show the potential effect of such measures.

And last but not least the instrument of public procurement can favour the use of biolubricants. An important condition for effective government interventions is that she sets the right example herself. In the case of biolubs public bodies and authorities make a considerable part of lubricant use on and around inland and coastal waters. When public bodies seriously attempt to alter their own lubricant use, their authority increases to require other users to do the same.

Finally it can be concluded that the chosen innovation approach, possible within the Innovation Programme, with a strongly market oriented content focussing on the identification of drivers and barriers and trying to formulate integrated answers to all these obstacles, has resulted in a strong awareness about the existence and benefits of biolubricants in environmental and technical terms, and it was a stimulus for the industry to continue and even intensify its R&D and marketing activities to develop and sell biolubricants.

Now it's the national and European governments' turn as well to adopt the conclusions and transform them in national and European interventions, stimulating measures and legislation.

# ANNEXES

# LLINCWA

## Action plan

## Market Segments

*Hydroelectric power*
*Inland Shipping*
*Water Management*
*Recreation*

**July 2001**
**Amsterdam/Gembloux/Toulouse/Hamburg/Eibar/Paris**

**Chemiewinkel UvA**
**Valonal**
**INPT**
**ISSUS**
**Tekniker**

# Contents

# Guidelines for filling out the chart

1) **The priority actor** : the priority actor is our ultimate target. This person , set of people, or organisation is the one who the project partners will target in order to get their support.
2) **The person responsible** : is the person within the project members that will be responsible for this segment of the action plan: contacting the priority actors, study their needs, reach them through LLINCWA's most adapted tools, respecting the timing and the budget.
3) **Short description of the priority actor** : who the target will be specifically?
4) **The needs of the actors** : is defined as the way the product responds to the specific needs of the target. What aspect of the product interests and responds to actor's needs?
5) **Types of influence the actor has on the project** : the influence an actor has on the project can be defined as* :
   **Legislation** : if the actor can through legislation encourage the use of biolubs, typically a member of the government or of the public administration
   **Communication** : if this actor is an entry point for spreading the knowledge the public has about biolubs
   **User** : the user of biolubs can have a positive impact on the diffusion of the product if he is properly targeted. That is if the actor is convinced of the effectiveness of biolubs he can "spread the word" and affect the product positively
   **Buyers** : people who buy the product
   **Prescribers** : The person or the group of people who have an influence on the choice or products the user/government/Buyer/etc can have
6) **The best way to reach them** : exemplifies the ideal way to reach the priority target. In this column all the ways to reach them will be enlisted. This should be very specific and clearly stated.
7) **The specific action for the project** : here the project partners will enlist not only the best way to reach the priority targets and the most adapted for this specific target, but also those that are possible and within the actual reach for LLINCWA members
8) **The material needed** : is the list of communication material that will be used to reach the priority target.
9) **Links to other countries** : is the list of countries that within the project can be of help or can co-operate in order to reach this target

## Market Segment  *Hydroelectric power*

### Spain / France

| | | |
|---|---|---|
| **Application:** | ES - | Hydraulic, gear oils |
| | F - | Grease and hydraulic fluids |

| | | |
|---|---|---|
| **Priority actor :** | ES - | person responsible for controlling the oil in different companies |
| | F - | Public company for electricity production, hydraulic system manufacturer and fitter |

| | |
|---|---|
| **Message :** | Bio lubricants help enhance the image of your company and improve the life of your equipment |

# Spain    Hydroelectric power
Hydraulics and gear oils

| Country | Short description of the priority actors | Needs of the actors | Type of influence the actor has on the project | Best way to reach them |
|---------|------------------------------------------|---------------------|------------------------------------------------|-------------------------|
| Spain | - Responsible for maintenance of equipments | -To increase the life of the lubricant.<br>- To reduce maintenance cost.<br>- To reduce the cost of the treatment of the used oil.<br>- To enhance the performance of the equipment, lower wear, reduce consumption<br>-To respect legislation and the environment | -User | - Distribution of the leaflets (2000)<br>- Distribution of the Newsletters (2000)<br>- Organisation of a workshop |

| Specific action for the project | Timing | Material needed | Person responsible | Links to other countries/segments | Questions to be solved |
|---------------------------------|--------|-----------------|--------------------|-----------------------------------|------------------------|
| -Distributing newsletter<br>-Direct mailing<br>-Distribution of brochures<br>-To send a questionnaire | May 01 | -Pilot demonstration projects (different manufacturers.)<br>-Newsletter, brochures, articles, etc | | France<br>Belgium<br>Netherlands<br>Germany | -Pilot projects data<br>-Other Courses that can be jointly uses to spread LLINCWA initiative.<br>-Finding out the number of shipping companies and hydroelectric companies non customers (phone numbers, names, attitude) |
| -Organisation of a workshop. | May 01 | | | | |
| -Distributing newsletter<br>-Direct mailing<br>-Distribution of brochures | May 02 | | | | |
| Include information about LLINCWA in the leaflets that TEKNIKER uses to create monthly. | May 02 | | | | |

5

# France,    Hydroelectric power
Grease and hydraulic fluids

| Country | Short description of the priority actors | Needs of the actors | Type of influence the actor has on the project | Best way to reach them |
|---|---|---|---|---|
| France | public company (EDF : electricité de France), the manufacturer (ex : CT OI), the fitter (ex: SHEM : société d' hydro-électricité du Midi) | To have technically performing products<br><br>To have information about environmental friendly products | the choice of the nature of lubricant in the considered area | Sailing campaign: communication on this topic through poster and workshop<br><br>To provide suitable information (technical and environmental data, references) |

| Specific action for the project | Timing | Material needed | Person responsible | Links to other countries/segments | Questions to be solved |
|---|---|---|---|---|---|
| Send invitations to the workshop during the sailing campaign<br>Distribute information brochure | June 01 | OK<br><br>Special brochure | | | Who write the technical book ? |

# Market Segment    *Inland Shipping*

## The Netherlands                **Inland Shipping OEMs**
**Closed Stern tube**

**Application:**       NL- closed stern tube

**Priority actor :**   NL- OEM's (Stern tube manufacturers)

**Message :**          The use of biolubs in closed stern tubes is a field-proven technique, which apart
                       from being more environmentally sound, it is cost neutral when compared to the
                       use of mineral lubricants in closed stern tubes

| Country | Short description of the priority actors | Needs of the actors | Type of influence the actor has on the project | Best way to reach them |
|---|---|---|---|---|
| NL | In the Netherlands six stern tube manufacturers are active in the market:<br>- De Waal<br>- IHC Lagersmit<br>- Mapron Engineering<br>- John Crane/Lips<br>- Amartech<br>- Uittenboogaard<br><br>The first three companies together supply more than eighty percent of the professional Dutch shipping market. | - Passive<br><br>Information on technical performance and cost aspects. | -They advice and give guarantees on products that may be used in their systems. The OEM makes the advised/ guaranteed products available in the so-called list of approval | Direct e-mail or via suppliers of biolubs.<br><br>_____<br><br>Workshop during the sailing campaign (the OEM's should be addressed together with the skippers) |

| Specific action for the project | Timing | Material needed | Person responsible | Links to other countries/segments | Questions to be solved |
|---|---|---|---|---|---|
| -- Set up projects<br><br>Gather information on ongoing tests conducted by the OEM's. | Throughout the LLINCWA project | Contentment evaluation check list<br><br>**Protocol for the performance of measurements during tests.**<br><br>LLINCWA folder | ori | France<br>Belgium<br>Germany | -Is there a need for the development of a sector specific folder / booklet |
| Make a plan of action for the workshop. | May 2001 | LLINCWA newsletters | | | |
| Workshop | October 2001 | | | | |

# The Netherlands                    Inland Shipping Government
open and closed stern tube

**Application:**          open and closed stern tube

**Priority actors:**      Government

**Message:**              Biolubs in open as well as in closed stern tubes are the environmentally sound
                          alternative for mineral products. Promoting biolubs by regulatory endorsements
                          responds to the national environmental policy concerning the tackling of diffuse
                          sources of water pollution. No change can be achieved without legislative support
                          (especially on the field of open stern tubes).

| Short description of the priority actors | Positioning/ Needs of the actors | Type of influence the actor has on the project | Best way to reach them |
|---|---|---|---|
| International<br>- CCNR[1]<br>- ICPR[2]<br>- EU<br>- International Commission for the Protection of the Danube | Passive<br><br>Information on all aspects (technical, environmental, costs) of the use of biolubs | Policy makers | - CCNR via the Dutch Commissioner<br>- ICPR via via the Dutch Commissioner in the CCNR<br>- The Danube commission will be addressed via the CCNR |
| National NL<br><br>- Ministry of Transport, Public Works and Water Management<br>- Ministry of Environment<br>- Working group surveillance Inland Shipping (Water police, Rotterdam Harbour, DG Public Works, Water Management (RWS)<br>- Water management authorities (water boards, Rijkswaterstaat, Province) | | | Via representatives of these authorities in the Dutch advice group of the project |

| Specific action for the project | Timing | Material needed | Links to other countries/segments | Questions to be solved |
|---|---|---|---|---|
| Organise workshop/ lecture during sailing campaign | - CCNR on 1/10/01<br>- ICPR on 8/10/01<br>- EC on 21/9/01<br>- EP 2/10/01 | LLINCWA folder<br><br>LLLINCWA news briefs<br><br>Pilot projects results | Germany, Belgium, France<br><br>The meeting with EC will be organised by Valonal (BE)<br><br>The meeting with EP will be organised in cooperation with INPT (France) | |

---

[1] Central Commission for the Navigation on the Rhine = international legislator for the Rhine
[2] International Commission for the Protection of the Rhine = international water manager for the Rhine

# Inland Shipping Netherlands / France / Belgium / Germany

**Application:** NL- open and closed stern tube[3]

F- hydraulic circuits, generating sets, stern tube

D- Hydraulic fluids, cables, stern tubes

B- stern tube

**Priority actor :** NL- shipping companies (big, medium and family owned companies)

F- Shippers, inland shipping hauliers

D- Ship owners, national inland shipping associations, ship suppliers, Universities or institutions that deal with ship related activities

B skippers and future skippers

**Message :** NL The use of biolubs in open stern tubes is a field-proven technique that can be immediately applied without any adjustment in the existing systems while leading to a substantial reduction of the environmental load of the surface waters stemming from inland shipping. The higher operational costs related to the higher price of biolubs can be (partly) compensated by the improved environmental image gained in this way that enhances the shipping company's opportunities to do business with authorities and other companies committed to implementing environmental standards (like the ISO 14000). Moreover, an improved environmental image can lead to a reduction of environmental fines and time spent on compliance issues. The use of biolubs in **closed stern tubes** is a field-proven technique, which apart from being more environmentally sound, it is cost neutral when compared to the use of mineral lubricants in closed stern tubes/

F. To contribute to the preservation of the water quality chose biodegradable oils in hydraulic circuits, and biodegradable grease on stern tube /

B To contribute to the preservation of the water quality, you can chose biodegradable grease on stern tube, as this kind of lubrication is totally lost in water

---

[3] There are two major applications of lubricants on-board: engine oils and stern tube lubricants. Other on-board applications have a lot smaller use volume, like the lubrication of the rudders, the anchor and the robes. The engine oils represent the biggest volume, but is not a loss or high-risk loss lubricant. If the oil leaks out of the engine, it will not spread diffusely and will not enter the water phase. In the Netherlands approximately 6000 ships are operational. About seventy percent still has a (half-) open grease lubricated stern tube. Almost all of them are filled with mineral oil based grease. Of the remaining thirty percent about ten percent is water lubricated and three quarters have an closed oil or grease lubricated system. In about ten percent of the closed stern tubes biolubs are already used.

# The Netherlands        Inland shipping
open and closed stern tube

| Country | Short description of the priority actors | Needs of the actors | Type of influence the actor has on the project | Best way to reach them |
|---|---|---|---|---|
| NL | The majority of the employees occupied in inland shipping work for themselves or in a family-owned business | Passive<br><br>Information on technical performance and cost aspects. | -Users | Via suppliers of biolubs[4]<br><br>Workshops / lectures for inland shipping associations to be hold preferably at their annual meetings[5]<br><br>Publishing in specialist / professional magazines[6]<br><br>Fairs[7]<br><br>Direct mailing<br><br>Ship shops<br><br>Working during the sailing campaign |

---

[4] Suppliers of biolubs active in this sector in the Netherlands

| Company name | Contact person | Tel (+31) |
|---|---|---|
| TotalFinaElf | | |
| Exxon Mobile | | |
| Shell | | |
| Castrol | | |
| Oman Benelux | | |
| Fuchs | | |
| Mecacyl Nederland | | |

[5] Two inland shipping associations have been identified in the Netherlands:
- CBRB (big shipping companies, 4% of skippers)
- Het Kantoor (small companies, 21% of skippers)

[6] There are several magazines of importance:
- Schuttevaer is a widely-read magazine. It appears weekly.
- Scheepvaartkrant
- Magazine Binnenvaart is published by the Netherlands association of (small and medium sized) shipping companies: Het Kantor
- Rijn & Binnenvaart is published by the Netherlands association of (big) shipping companies: CBRB
- Binnenvaartkrant

[7] The most important fairs for this sector are the Rotterdam Maritime (Ahoy, november 2002) Europoort (RAI, november 2001) en Duisburg (september 2001)

| Specific action for the project | Timing | Material needed | Person responsible | Links to other countries/ segments | Questions to be solved |
|---|---|---|---|---|---|
| - Set up projects<br><br>- Gather information on ongoing tests conducted by the suppliers. | Throughout the LLINCWA project | Contentment evaluation check list<br>Protocol for the performance of measurements during tests.<br>----------------- | | France<br>Belgium<br>Germany | -Is there a need for the development of a sector specific folder / booklet |
| Establish contacts with inland shipping associations. | 07/ 01 | -LLINCWA folder<br>-LLINCWA newsletters<br><br>Articles (3x) | | | |
| Press release<br>Write articles<br>Establish contacts with editors | 07/ 01<br>04/ 02<br>12/ 02<br>05/ 01 | | | The articles could be used for the Flemish market | Which magazines should be approached? |
| Visit Rotterdam Martitime<br>Visit Europoort<br>Visit Duisburg<br>Disseminate information<br><br>E-mail | -Ahoy 11/02<br>-Rai 11/01<br>-10/01 | Posters<br>LLINCWA folder &Newsletters | | All partner countries | Investigate the possibility of organising a seminar during the fair<br>Is there a need for the development of a sector specific folder / booklet?<br><br>Is there a need for the development of a sector specific folder / booklet? |
| Make a list of important ship-shops<br>Mailing<br>Visits | All during the project<br><br>05/01<br>07/01<br>During the sailing campaign | LLINCWA Folder &Newsletters | | | |
| Make a plan of action for the workshop<br>Workshop | 05/01<br><br>10/01 | | | Flemish skippers may attend the workshop | |

**Belgium**                    **Inland Shipping**
                               Stern tube

| Country | Short description of the priority actors | Needs of the actors | Type of influence the actor has on the project | Best way to reach them |
|---|---|---|---|---|
| Belgium | Mostly individuals skippers, represented in some branch associations | Technical data and prices on products<br><br>Info and guarantee from OEM and supplier | They choose the type of stern tube grease | Publicity in press for inland shipping (" TB-info" distributed the skipper by ITB (Institut pour le Transpot par Batellerie)<br><br>Inland shipping schools[8] |

| Specific action for the project | Timing | Material needed | Person responsible | Links to other countries/segments | Questions to be solved |
|---|---|---|---|---|---|
| Newsletters (and leaflet) could be attached to the current "ITB-info"<br><br>I propose that the Flemish skippers receive the tools coming from NL<br><br>Visit of the Reinwater exhibition | 04/01<br>09/01<br>04/02<br><br><br><br><br>Sailing campaign 2 (5/02) | Newsletters 1 (OK), 2, 3<br><br><br><br><br><br>Posters and C° on board of Reinwater (OK) | | With NL for the tools to target the skippers of the Flemish Region | |

---

[8] Koninklijk Technisch Atneneum "Ruggeveld", Antwerpen
Section "Centflumarin", Zwijndrecht-Kallo
Centre de Formation en Alternance, Section batellerie, Huy

# Germany    Inland Shipping
Hydraulic fluids, cables, stern tubes

| Country | Short description of the priority actors | Needs of the actors | Type of influence the actor has on the project | Best way to reach them |
|---|---|---|---|---|
| Germany | Ship owners<br>National inland shipping associations<br>Ship suppliers<br>Technical universities/institutions for ship building/engineering/construction (e.g. Mercado university, Duisburg) | | | |

| Spec action | Tim ing | Mat | Pers resp | Links | Questions to be solved |
|---|---|---|---|---|---|
| | | | | | Information gathering<br>Per country it is essential to know:<br>-Which shipping fairies take place during the next two years?<br>-Which branch journals are relevant?<br>-What are the national shipping associations?<br>-Which technical universities are relevant?<br>-Who will be invited for the sailing campaign<br>-What is the attitude of ship owners towards biolubs?<br>-Success stories gained with publicly owned ships?<br>-Who is in charge within LLINCWA to manage the market segment?<br>-Invitation to the campaign |

# Market Segment : *Water Management*

## Belgium / France /Germany –Water Management

| | |
|---|---|
| **Application:** | Grease and hydraulic fluids |
| **Priority actor :** | **B-** Public managers of inland water and havens |
| | **F-** Managers of waterways (VNF) |
| | **D-** Lower Water Authority |
| | **ES-** Maintenance Equipments |
| **Message :** | Contribute to the preservation of water quality by choosing biodegradable lubricants |

# Belgium           Water management
## Grease and Hydraulic fluids

| Country | Short description of the priority actors | Needs of the actors | Type of influence the actor has on the project | Best way to reach them |
|---|---|---|---|---|
| Belgium | Managers of the waterway (who used to command product they need to run locks, mobile bridges, dams, etc).[9] | To have technically performing products<br><br>To have infos on products environmentally friendlier | Depending on personal conviction, someone can chose a biolub in place of non-bio product | <u>Publicity in Branch journals</u> ("Les Infos du MET" in Walloon Region is distributed to all the people working in the public body MET (Ministry for Equipment and Transport)<br><u>Sailing campaign</u>: to promote biodegradable greases and hydraulic oil give some technical facts in newsletters led by the locks that the cruise meet during its travel<br>Make specific report on technical aspects on grease and hydraulics applied to locks, dams, mobile bridges, … and sent it to the managers of the public bodies<br>Workshop with people working for public water management on board of the Reinwater |

| Specific action for the project | Timing | Material needed | Person responsible | Links to other countries/segments | Questions to be solved |
|---|---|---|---|---|---|
| - Newsletters (and leaflet) will be attached to the current "Les Infos du MET"<br>- I propose that the Flemish public bodies receive the tools coming from NL<br>- Make and distribute the technical report<br>-Invitation workshop during sailing campaign | 04/01<br>09/01<br>04/02<br><br>2nd sailing campaign (may 2002)<br><br><br>asap<br><br>Before 4/02 | Newsletters 1 (OK), 2, 3<br><br>General leaflet (OK)<br><br><br><br>Technical book | | With NL for the tools to target the public body of the Flemish Region | Who write the technical book ?<br><br>Find the exact names of the public body in Flemish Region |

---

[9] For Walloon Region : MRW/MET/Division Services techniques and Division Voies Navigables,Port de Bruxelles
For Flemish Region : Ministerie van Vlaamse Gemeenschap A.W.Z.
Dienst voor de Scheepvaart, Hasselt
Zeekanaal N.V. , Willebroek
M.B.Z. , Brugge

# Germany          **Water management**
Grease and Hydraulic fluids

| Country | Short description of the priority actors | Needs of the actors | Type of influence the actor has on the project | Best way to reach them |
|---------|-------------------------------------------|---------------------|------------------------------------------------|------------------------|
| Germany | | | Buyers first example legal | |

| Specific action for the project | Timing | Material needed | Person responsible | Links to other countries/segments | Questions to be solved |
|---------------------------------|--------|-----------------|--------------------|-----------------------------------|------------------------|
| - Half day exchange of experiences during the Sailing campaign (each stop)<br>- Lobbying<br>- Internal magazines<br>- Final conference | | -March 01<br>- June 01<br>- Sept 01 | | | |

# France      Water Management
Grease and Hydraulic fluids

| Country | Short description of the priority actors | Needs of the actors | Type of influence the actor has on the project | Best way to reach them |
|---|---|---|---|---|
| France | Managers of the waterway (who are used commanding lubricants to run locks, mobile bridges, dams, etc). | To have technically performing products<br><br>To have infos on products environmentally friendlier | The choice of the nature of lubricant in the considered area | Publicity in Branch journals: internal journal for VNF, journal distributed by DIREN<br><br>Sailing campaign: communication on this topic through poster and workshop<br><br>Make a specific leaflet meant for VNF network |

| Specific action for the project | Timing | Material needed | Person responsible | Links to other countries/segments | Questions to be solved |
|---|---|---|---|---|---|
| Leaflets, newsletter can be attached to internal revues | May-June 2001 | OK | | | Who write the technical book ? |
| Distribute technical informations | Sailing campaign | Technical report | | | |
| Send invitations to the workshop during the sailing campaign | June 2001 | | | | |

**Spain**          **Water Management**

Grease and Hydraulic fluids

| Country | Short description of the priority actors | Needs of the actors | Type of influence the actor has on the project | Best way to reach them |
|---|---|---|---|---|
| Spain | - Responsible for maintenance of equipments | -To increase the life of the lubricant.<br>- To reduce maintenance cost.<br>- To reduce the cost of the treatment of the used oil.<br>- To enhance the performance of the equipment, lower wear, reduce consumption<br>-To respect legislation and the environment | -User | - Distribution of the leaflets (2000)<br>- Distribution of the Newsletters (2000)<br>- Organisation of a workshop |

| Specific action for the project | Timing | Material needed | Person responsible | Links to other countries/segments | Questions to be solved |
|---|---|---|---|---|---|
| -Distributing newsletter<br>-Direct mailing<br>-Distribution of brochures<br>-To send a questionnaire<br><br>-Organisation of a workshop.<br><br>-Distributing newsletter<br>-Direct mailing<br>-Distribution of brochures<br><br>Include info on LLINCWA in the leaflets TEKNIKER uses to create monthly. | May 2001<br><br>May 2001<br><br>May 2002<br><br>May 2002 | -Pilot demonstration projects (different manufacturers.)<br>-Newsletter, brochures, articles, etc | | France<br>Belgium<br>Netherlands<br>Germany | -Pilot projects data<br>-Other Courses that can be jointly uses to spread LLINCWA initiative.<br>-Finding out the number of maintenance responsible of water management organisations non customers (phone numbers, names, attitude) |

18

# Belgium/France – market segment Water Management

**Application:** B-F-NL Grease and hydraulic fluids
NL gear box oil
**Priority actor :** B – Deciders involved in Public body purchasing
F- Water agency, water producer company
NL- Water boards, National Water Management Authority
**Message :** To contribute to the preservation of water quality, propose biodegradable lubricants
for the (industrial) applications near water

## Belgium    Water management
Grease and Hydraulic fluids

| Country | Short description of the priority actors | Needs of the actors | Type of influence the actor has on the project | Best way to reach them |
|---|---|---|---|---|
| Belgium | Managers of the public sector who are in charge of the public purchasing, (they edit a list of product) | To have technical data and prices on products (cost comparison)<br><br>To have an information on environmental issues of lubricants | They can put biolubs on the approval list distributed to managers | -Publicity in Branch journals ("Les Infos du MET" in Walloon Region is distributed to all the people working in the public body MET (Ministry for Equipment and Transport),<br>-Make specific report on economical (and technical aspects on grease and hydraulics applied to locks, dams, mobile bridges, … and sent it to the managers of the public bodies<br>-Reinwater – Workshop<br>-Direct contact with the specific desk |

| Specific action for the project | Timing | Material needed | Person responsible | Links to other countries/segments | Questions to be solved |
|---|---|---|---|---|---|
| Newsletters (and leaflet) will be attached to the current "Les Infos du MET"<br><br>I propose that the Flemish public bodies receive the tools coming from NL<br><br>Invitation to a workshop during the  sailing campaign<br>Make and distribute the technical report | 2$^{nd}$ sailing campaign (may 2002)<br><br>-04/01<br>-09/01<br>-04/02<br>-Before 4/02 | Newsletters 1 (OK), 2, 3<br><br>Technical book | | With NL for the tools to target the public body of the Flemish Region | Who write the technical book ?<br><br>Find the exact names of the public body involved in purchasing in Flemish Region |

**France**          **Water management**

Grease and Hydraulic fluids

| Country | Short description of the priority actors | Needs of the actors | Type of influence the actor has on the project | Best way to reach them |
|---------|------------------------------------------|---------------------|------------------------------------------------|------------------------|
| France | -Managers of the public sector in charge of water quality control [10] - managers of private sector in charge of producing drinking water[11] | To have technical data and prices on products (cost comparison) To have an information on environmental issues of lubricants | they can act in favour of biolubricants : put biolubs on approval list, define clauses including biolubs (2) or incitatives (1) | Publicity in Branch journals: internal journal of water agency and journal edited by water company Make specific report on economical and technical aspects on grease and hydraulics substitution Direct contact with the specific desk |

| Specific action for the project | Timing | Material needed | Person responsible | Links to other countries/segments | Questions to be solved |
|---------------------------------|--------|-----------------|--------------------|-----------------------------------|------------------------|
| Newsletters (and leaflet) will be attached to the journals Invitation to a workshop during the sailing campaign | June 2001 | OK | | gather the technical data | Who write the technical book ? |
| Distribute the technical report | Second sailing campain asap | Technical book | | | |

---

[10] Agence de l'eau, DRIRE
[11] Vivendi, Lyonnaise de l'eau

# Netherlands    Water management
Gear Box oil, Grease and Hydraulic fluids

| Country | Short description of the priority actors | Needs of the actors | Type of influence of actor on the project | Best way to reach them |
|---------|------------------------------------------|---------------------|-------------------------------------------|------------------------|
| Holland | **Water boards** (local government bodies that manage the quality and quantity of water in their area **Min. Transport Public Works and Water Management** | Information on technical performance of biolubs Information on cost aspects | -Users -Policy makers (indirect in the case of water boards)[12] | Via suppliers of biolubs [13] Publishing in specialist magazines[14] Sustainable purchasing program bureau [15] Fairs [16] Direct mailing Workshop during the sailing campaign Via the association of water boards Via the platform to diffuse sources of pollution [17] |

---

[12] Higher levels of government consult water boards for the development of national policy. Besides, water boards draw up a management plan for local and regional waters according to objectives and instructions of the higher level of government.

[13] Suppliers of biolubs active in this sector

| Company name | Contact person | Tel (+31) |
|--------------|----------------|-----------|
| Shell | | |
| Castrol | | |
| Exxon Mobil | | |
| Texaco Rotterdam | | |
| Kuwait Petroleum Rotterdam | | |
| BP  Rotterdam | | |

[14] There are three magazines of importance:
- **Het Waterschap** is a widely- read internal magazine for administrators of water boards. The magazine appears once every three months. The person who should be addressed for placing an article about LLINCWA in this magazine is Arno van Bremen from the Association of water boards
- **H₂O** is the official magazine of several Dutch associations for water management and waterworks ( NVA, VEWIN, VWN and Kiwa). It appears once every fourteen days. It is well read by technicians and engineers.
- **Neerslag** is published by NVA (Nederlandse Vereniging voor Waterbeheer / Nederlands Association for Water Management) for its regional divisions. It appears once every two months.

[15] The sustainable purchasing program is a governmental initiative that encourages and assists governmental authorities in buying environmentally preferred products. By listing biolubs as defined within LLINCWA as environmentally preferable products promoted by the sustainable purchasing program bureau stimulates the use of biolubs by water management authorities committed to the sustainable purchasing program. At the moment 15 out of 57 water boards have committed themselves to this program.

[16] The most important fair for this sector is Aquatech. Aquatech is an international trade exhibition of water technology and water management

[17] The platform of diffuse sources pollution is consultative structure between the water management and environmental authorities on regional level aimed at the tackling of pollution derived from diffuse sources.

| Specific action for the project | Timing | Material needed | Person responsible | Links to other countries/segments | Questions to be solved |
|---|---|---|---|---|---|
| Set up the projects | All through out the LLINCWA project | Contentment evaluation check list -Protocol for the performance of measurements during tests -LLINCWA folder &Neswletter | | | Is there a need for the development of a sector specific folder / booklet? |
| Write articles  Establish contacts with editors | 7/ 2001 4/ 2002 12/2002 5/ 2001 | Articles (3x) | | The articles could be used also for the Flemish market. | Which magazines to approach? |
| Mailing  Personal contact | 2/ 2001  5/ 2001 | Ranking system document 1ste newsletter folder | | | |
| Visit Aquatech Amsterdam RAI  Disseminate information | 1-4 /10/02 | Posters folder &Neswletter | | This activity is interesting for all the partners deploying activities in the water managment sector | Investigate the possibility of organizing a seminar during this fair. -Is there a need for the development of a sector specific folder / booklet? |
| | During entire project | folder & Newsletter | | | |
| - Make an action plan for the workshop - Workshop | May 2001  Oct /01 | idem | | Flemish water authorities can attend the WS | |
| Personal contact Attending the meeting | 03/01 04/01 | idem | | | |
| Mailing to the regional platforms | 05/01 | idem | | | |

## Market Segment : *Recreation*

### Belgium / France – market segment : RECREATION

**Market segment :**  Recreation
**Application:**  **B/F**  Two stroke for outboard motors
  **NL**  + stern tube
**Priority actor :**  Yachtsmen /professional /individual users /
**Message :**  To contribute to the preservation of the water quality, you can chose
  biodegradable lubricants for most of the applications  near water

### Belgium  Recreation
Two stroke for outboard motors

| Country | Short description priority actors | Needs of the actors | Type of influence the actor has on the project | Best way to reach them |
|---|---|---|---|---|
| Belgium | Yachtsmen, water-sports men, tour operators with boats tours, resellers of oils for yachting | To enjoy their hobby without destroy the quality of the water | Individual decision, information | - Publicity in Branch journals[18]<br>- Make a specific leaflet to promote biodegradable two-stroke oil and disseminate it to the locks and harbour managers; ask them to distribute to pleasure boats skippers.<br>- Diffusion leaflets by associations and resellers<br>- Specific fairs (Salon des vacances, Foires nautiques, …):  contact the managers, insert a message in the program paper for example. |

| Specific action for the project | Timing | Material needed | Person responsible | Questions to be solved | Links to other countries/segments |
|---|---|---|---|---|---|
| Send press release to branch journals | | Newsletter (general) | | Who write the leaflet ? | |
| Disseminate specific leaflet through harbour managers, Promotions Offices, specific associations,… | | Specific leaflet on 2-stroke oil | | Are there pilots on 2 stroke oil ? | |

---

[18] pleasure boats associations, aquatic sports associations

## France          ## Recreation

Two stroke for outboard motors

| Country | Short description of the priority actors | Needs of the actors | Type of influence the actor has on the project | Best way to reach them |
|---|---|---|---|---|
| France | Yachtsmen, water-sports men, tour operators with boats tours, resellers of oils for yachting, fluvial policy, rescue brigade, maintenance patrol | To preserve the medium where they practice their activities | Individual decision, information | Publicity in professional journals<br><br>Make a specific leaflet to promote biodegradable two-stroke oil for dissemination<br><br>Specific fairs or manifestations: contact the managers, insert a message in the program paper for example. |

| Specific action for the project | Timing | Material needed | Person responsible | Links to other countries/segments | Questions to be solved |
|---|---|---|---|---|---|
| Send articles to professional journals<br><br>Disseminate specific leaflet through harbour managers, leisure centre Promotions through information Offices, specific associations,… | end of 2001 | Specific leaflet on 2-stroke oil | Christine Ceccutti | | Who write the leaflet ?<br><br>Are there pilots on 2 stroke oil ? |

# Belgium/France – market segment RECREATION

**Application:** Two stroke oil for outboard motors
**Priority actor :** Public promotion bodies, managers of pleasure harbours and lakes, tourism organisations,... (around users)
**NL** + Co-ordination committee on environment, Tourism and Leisure (CETL) / owners
**Message :** The performance of biolubs and the quality of the product allow you to enhance the environmental quality of your tourists place, and will let you attract tourism and promote your site

| Country | Short description of the priority actors | Needs of the actors | Type of influence the actor has on the project | Best way to reach them |
|---------|------------------------------------------|---------------------|-----------------------------------------------|-------------------------|
| Belgium | Tour operators near water, public promotion bodies[19], tourism information and promotion bodies, natural parks and sensible areas, managers of pleasure harbour and lakes | To enhance the quality of a potential tourists place | Legislation, rules, individual decision, information | Make a specific leaflet to promote biodegradable two-stroke oil and disseminate it to the targets<br><br>Lobbying on the actors concerned by the pleasure lakes and harbour management |

## Belgium         Recreation

Two stroke oil for outboard motors

| Specific action for the project | Timing | Material needed | Person responsible | Links to other countries/segments | Questions to be solved |
|---------------------------------|--------|-----------------|--------------------|-----------------------------------|------------------------|
| Disseminate specific leaflet through harbour managers, Promotions Offices, and some tourists areas[20] | | Newsletter (general)<br><br>Specific leaflet on 2-stroke oil | MHN | | Who write the leaflet?<br>Are there pilots on 2 stroke oil ? What's the experience we want to show ? |

---

[19] Office de Promotion des Voies Navigables, Promotie Binnenvaart Vlanderen
20 Barrages de l'eau d'Heure, Plan incliné de Ronquière (Parc naturel des collines), ...

# France          Recreation

Two stroke oil for outboard motors

| Country | Short description of the priority actors | Needs of the actors | Type of influence the actor has on the project | Best way to reach them |
|---|---|---|---|---|
| France | Tour operators near water, tourism information, natural parks and sensible areas, managers of pleasure harbour and lakes | To enhance the quality of a potential tourists place | Legislation, rules, individual decision, information | Make a specific leaflet to promote biodegradable two-stroke oil and disseminate it to the targets

Lobbying on the actors concerned by the pleasure lakes and harbour management |

| Specific action for the project | Timing | Material needed | Person responsible | Links to other countries/segments | Questions to be solved |
|---|---|---|---|---|---|
| Disseminate specific leaflet through harbour managers, Promotions Offices, and some tourists areas[21] | | Newsletter (general)

Specific leaflet on 2-stroke oil | | | Who write the leaflet ?

Are there pilots on 2 stroke oil ?  What is the experience we want to show ? |

---

...

# Netherlands          Recreation

Two stroke oil for outboard motors + stern tube

| Country | Short description of the priority actors | Needs of the actors | Type of influence the actor has on the project | Best way to reach them |
|---|---|---|---|---|
| Holland | CETL is the national platform for sustainable development of tourism. It is private-public co-operation. It consists of representatives of the five Ministries, the Association of Provincial Authorities and twelve lead organisations representing the tourism & leisure sector in The Netherlands.<br><br>Owners | General information on the availability of products their performance and related cost aspects | Policy makers<br><br><br>Users | Attending meetings<br><br>Disseminate information<br><br><br>Disseminate information during faires (Hiswa) |

| Specific action for the project | Timing | Material needed | Person responsible | Links to other countries/segments | Questions to be solved |
|---|---|---|---|---|---|
| Write a specific folder /booklet for the sector | July 01 | LLINCWA folders<br><br>Newsletters Sector Specific folder | | | -When are the coming meeting of CETL?<br>-When is the next HISWA fair?<br>-Establish useful partnerships for the presentation of the LLINCWA message during fairs. |

| | | |
|---|---|---|
| IVAM, Chemical Risks<br>Postbox 18180<br>NL-1001 ZB  Amsterdam | Pieter van Broekhuizen<br>Demi Theodori<br>Hildo Krop<br>Ckees van Oijen | pvbroekhuizen@ivam.uva.nl<br>dtheodori@ivam.uva.nl<br>hkrop@ivam.uva.nl<br>coijen@ivam.uva.nl<br><br>tel: 0031 20 525 5080<br>fax: 0031 20 525 5850 |
| QA+, Questions Answers and More BV<br>P.O.Box 137<br>NL- 2501 CC  Den Haag | Kees Le Blansch | klb@qaplus.nl<br><br>Tel: 0031 70 30 25 838<br>Fax: 0031 70 30 25 839 |
| Hoogheemraadschap van Rijnland<br>Archimedesweg 1<br>P.O.Box 156<br>NL-2300 AD  Leiden | Cees Ouwehand | Couweh@hhrsrijnland.nl<br><br>Tel: 31 71 51 68 991<br>Fax: 31 71 51 23 916 |
| TotalFinaElf<br>16, Rue de la Republique – Puteaux<br>F - 92970  Paris – La Defense | Philippe Lanore<br>Bernard Lamy | philippe.lanore@totalfinaelf.com<br><br>Tel +33 1 41 35 8267<br>Fax:+33 1 41 35 8561 |
| Tekniker Research Foundation<br>Avda Otaola 20<br>E-20600  Eibar | Amaya Igartua | aigartua@tekniker.es<br><br>Tel: 0034 943 206 744<br>Fax: 0034 943 202 757 |
| INP ENST<br>Laboratoire de Chimie Agro-Industrielle<br>118 Route de Narbonne<br>FR-31077  Toulouse cedex 04 | Pascale de Caro<br>Christine Cecutti | psatge@ensct.fr<br>ccecutti@ensct.fr<br><br>Tel : 0033 5 62 88 57 25<br>Fax: 0033 5 62 88 57 30 |
| Fachhochschule Hamburg, ISSUS<br>Rainvilleterasse 4<br>DE-22765  Hamburg | Holger Watter | watter@issus.haw-hamburg.de<br><br>Tel: 0049 40 42875 6544<br>Fax: 0049 40 42875 6509 |
| Faculté universitaire des Sciences<br>agronomiques<br>Passage des déportés, 2<br>BE-5030  Gembloux | Marie-Hélène Novak | novak@valbiom.be<br><br>Tel: 0032 81 62 23 50<br>Fax : 0032 81 62 23 16 |
| Fuchs-Lubritech GmbH<br>Hans-Reiner-Strasse 7-13<br>DE- 67685  Weilerbach | Christian Busch<br>Uwe Schmidt | christian.busch@fuchs-lubritech.de<br>uwe.schmidt@fuchs-lubritech.de<br><br>Tel : 0049 6374 924 811<br>Fax : 0049 6374 924 774 |

T - #0268 - 101024 - C0 - 254/178/12 [14] - CB - 9789058096128 - Gloss Lamination